João Rodrigues Miguel
Marco Antonio Silva Vieira

Barão de Mauá Municipal Natural Park, Magé-RJ:

João Rodrigues Miguel
Marco Antonio Silva Vieira

Barão de Mauá Municipal Natural Park, Magé-RJ:

Pedagogical Space for Environmental Awareness

ScienciaScripts

Imprint

Any brand names and product names mentioned in this book are subject to trademark, brand or patent protection and are trademarks or registered trademarks of their respective holders. The use of brand names, product names, common names, trade names, product descriptions etc. even without a particular marking in this work is in no way to be construed to mean that such names may be regarded as unrestricted in respect of trademark and brand protection legislation and could thus be used by anyone.

Cover image: www.ingimage.com

This book is a translation from the original published under ISBN 978-613-9-66819-9.

Publisher:
Sciencia Scripts
is a trademark of
Dodo Books Indian Ocean Ltd. and OmniScriptum S.R.L publishing group

120 High Road, East Finchley, London, N2 9ED, United Kingdom
Str. Armeneasca 28/1, office 1, Chisinau MD-2012, Republic of Moldova, Europe
Printed at: see last page
ISBN: 978-620-8-06087-9

I dedicate this work to my fellow Master's students, for always encouraging and supporting me in this endeavour. To my loyal wife Dalva for sharing my achievements with me and for her understanding and patience in times of stress.

SUMMARY

CHAPTER 1	6
CHAPTER 2	8
CHAPTER 3	14
CHAPTER 4	15
CHAPTER 5	18
CHAPTER 6	21
CHAPTER 7	30
CHAPTER 8	37
CHAPTER 9	39

ACKNOWLEDGEMENTS

To God. To Him be all honour and glory!

To my family, wife, daughter, sons and mother, for always encouraging and supporting my choices and for understanding my absence at so many moments.

To my supervisor, João Rodrigues Miguel, for respecting my steps, for his sincere and loving welcome and for the constant encouragement that made me believe in my potential.

To the park guide, Adeilmanto Carlos da Silva, for his honourable, invaluable contribution to this work.

To the teachers at UNIGRANRIO, for their encouragement throughout the course and for enabling us to build quality knowledge.

To the students of the Hilda de Souza Coelho School, for welcoming me into their field classes with kindness and respect and for kindly giving me their participation so that I could reproduce them and include them in my research.

To the Magé Municipal Department of the Environment, for making the UC available so that I could carry out my research.

To my classmates, for sharing knowledge together and for the joyful and tense moments when we had to complete group work.

To everyone who was directly or indirectly involved in the realisation of this work.

Thank you very much!

"The. choice is ours: *form a* global alliance to *care for the earth and each* other, or risk our destruction and that of the diversity of life."

(Earth Charter, 1992).

SUMMARY

The aim of this research was to evaluate the application of interventions with 7th and 8th graders° at a public school in the Ipiranga neighbourhood, which is located in the mangrove biome, and to help raise their awareness of environmental issues through educational activities related to this biome. The environmental issues were presented to the students of the Hilda de Souza Coelho School. The methodological framework used was qualitative research, with the help of instruments such as a presentation of the study area, walks on the interpretive trails and the application of a questionnaire. Confirming aspects of the theory that the trails have the purpose of developing a new field of awareness in hikers, we found out how the students perceived the environment and the park's environmental problems. They developed an understanding that environmental interpretation, like environmental education, is not just about transmitting information, but about involving values, feelings and care for the space visited, seeking to reflect on the meaning of the environment. There is a need to reverse this situation of degradation and abandonment in favour of sustainable preservation, based on environmental education and raising environmental awareness of this natural heritage, whose main biome iu the mangrove swamp, which is of great importance not only to the conservation area, but also to the entire surrounding community. The final considerations conclude the dissertation, demonstrating a close interaction between the students and the activities carried out. The conclusion is that it is an important step in the students' education to include activities in natural areas in their environmental education projects.

Keywords: Environmental Education, Environmental Awareness, Interpretive Trails.

CHAPTER 1

I'm Marco Antônio da Silva Vieira, born in the city of Rio de Janeiro on 24 January 1964, son of Júlio Gomes Vieira and Maria Olinda da Silva, living in the city of São João de Meriti, in the Baixada Fluminense. Professionally, I work in the teaching team at FAETEC - Imbariê - Secondary School, as a Biology teacher.

The subject of this research was not chosen at random, nor was it the product of speeches by environmentalists who want to save the planet; it stems mainly from the many years I have worked as a primary and secondary school teacher. If, in my initial training, I chose to study Biological Sciences because of the need to understand social and environmental issues such as poverty, hunger, underdevelopment, environmental imbalance, deforestation and social inequality, my entry into teaching gradually permeated the paths that would lead me to discover how environmental issues and their different meanings exerted great influence and motivation in my classes.

Faced with these motivations and experiences, I felt the need to delve deeper into these issues and look for possible answers to understand the serious environmental problems that were highlighted in the daily news, in the course of the lessons and even on a local scale, from the different spaces inhabited and experienced by the students.

In 2005, I took a postgraduate course in the environment at the Cândido Mendes University. It was my first specific contact with the subject, and through this concept I sought a way of deepening my knowledge, believing it to be a space for reflection on the environmental issues that materialise within the existing reality.

Even on this journey, it wasn't possible to gain a deeper understanding of the meaning of Environmental Education (EE), because although there was a strong association between man and nature, I think that the readings I did had a great influence on my education and further reinforced my feeling that raising awareness would be an important strategy for changing the hegemonic thinking that sees the environment as a commodity capable of providing natural resources for the constant progress of capitalist society (Netto, 1980). I became increasingly aware that this apprehension of the sensitive and affective could lead society to a new relationship with nature.

But this view of the environment is gradually giving way to new conceptions, opening up new ways of looking at things and, consequently, deepening my theoretical framework on the subject. Even so, some authors continue to point out that raising awareness is the first step when you want to carry out environmental education actions (Tristão, 2004; Viegas, 2004), but that the search for affective and behavioural identity alone is insufficient to give visibility to and legitimise environmental education practices.

Considering this whole learning process, I was still torn between two areas that were strongly related to

Biology and which increasingly became topics of individual interest and motivations for my career as a teacher: environmental issues and those related to social problems. The realisation of my dream of studying for a master's degree was therefore postponed for many years, as I hesitated between the teaching or academic modality. Throughout these years, many discoveries were made until I arrived at the master's programme at Unigranrio University. When I learnt that I could investigate EE in schools, I decided to enter the selection process.

Environmentally correct behaviour must be learnt in practice, contributing to the formation of responsible citizens. In this context, there is nothing more significant and important than starting to raise environmental awareness by introducing environmental education in a primary school, so that today's children can become part of fairer societies in the future, where citizens respect themselves, their fellow human beings and all forms of life in the environment. It is therefore necessary to set up projects aimed at raising awareness among students and other residents living in the vicinity of a PA with a diversity of biomes, valuing their environmental aspects and giving due importance to the environment. The aim of environmental education is to sensitise individuals to the environment in which they live, so that they can enjoy a better quality of life while respecting the nature that surrounds them (Mansano, 2006).

As a Biology and Science teacher, it is because I have a great interest in environmental education and because the area of research has had a great influence on my life, that the desire arose to try to understand the importance of the environmental issue in the context of the lives of students living around a Conservation Unit (CU). Using its natural attributes, I tried to understand whether the environment and its interactions occupied any space in the daily lives of these students, or whether they were considered unimportant factors. The initial idea of working with the theme "Interpretive Trails in a Non-Formal Space" gradually took on a very significant dimension, as the research was redesigned and other paths were travelled.

We believe that environmental education is the basis for initiating a new socio-environmental change, enabling interdisciplinarity in whatever part of society it takes place, thus provoking transformations in the individual's awareness and sensitisation.

In this way, it was only fitting that the work was carried out in a school in the park's surroundings, with the intention of finding out from the students to what extent they would show all their sensitivity to the peculiar characteristics that they were able to contemplate, admire and recognise, giving importance to their local environmental, landscape and cultural value, feeling "special" for living in an environment with such enchanting images.

CHAPTER 2

THEORETICAL FRAMEWORK

2.1. Environmental Education in Brazil

Environmental Education (EE) emerged in Brazil long before it was institutionalised by the Federal Government. We have as a factor a persistent conservationist movement until the early 1970s, when an environmentalism emerged that joined the struggles for democratic freedoms, manifested through the isolated action of teachers, students and schools, through small actions by civil society organisations, city halls and state governments, with educational activities aimed at restoring, conserving and improving the environment. During this period, the first specialisation courses in environmental education also emerged (Brasil, 1999).

The process of institutionalising environmental education by the Brazilian federal government began in 1973 with the creation of the Special Secretariat for the Environment (SEMA), linked to the Presidency of the Republic. Another step towards institutionalisation was taken in 1981, with the National Environmental Policy (PNMA), which established, in the legislative sphere, the need to include environmental education at all levels of education, including community education, with the aim of empowering the community to actively participate in defending the environment. Reinforcing this trend, the Federal Constitution in 1988 established, in item VI of article 225, the "need to promote environmental education at all levels of education through public awareness for the preservation of the environment" (Brasil, 1999).

In December 1994, as a result of the 1988 Federal Constitution and the international commitments made during Rio-92, the Presidency of the Republic created the National Environmental Education Programme (PRONEA), shared by the then Ministry of the Environment, Water Resources and the Legal Amazon and the Ministry of Education and Sports, with partnerships from the Ministry of Culture and the Ministry of Science and Technology. PRONEA was implemented by the MEC's Environmental Education Coordination and the corresponding sectors of the MMA/IBAMA, responsible for actions aimed at the education system and environmental management respectively, although other public and private organisations in the country were also involved in its implementation (Brasil, 1999).

In 1995, the Temporary Technical Chamber for Environmental Education was created within the National Environmental Council (CONAMA). The guiding principles for the work of this Chamber were participation, decentralisation, recognition of cultural plurality and diversity, and interdisciplinarity (Brasil, 1999).

In 1996, the Environmental Education Working Group was set up within the MMA, and a protocol of intentions was signed with the MEC, aimed at technical and institutional co-operation, becoming a formal channel for the development of joint actions. After two years of debate, the National Curriculum Parameters (PCN) were approved by the National Education Council in 1997. The PCNs are a subsidy to support schools in drawing up their educational project, including procedures, attitudes and values in school life, as well as the need to deal with some urgent social issues of national scope, known as cross-cutting themes: the environment, ethics, cultural plurality, sexual orientation, work and consumption, with the possibility of schools and/or communities choosing others of relevant importance to their reality (Brasil, 1999).

In 2000, environmental education was included in the Multiannual Plan (2000-2003) for the second time, now as a Programme, identified as 0052 and institutionally linked to the Ministry of the Environment (Brazil, 2014).

In 2004, a new Multiannual Plan began, the PPA 2004-2007. As a result of the new guidelines in line with PRONEA, Programme 0052 was reformulated and renamed EE for Sustainable Societies (Brasil, 2014).

In 2007, the Chico Mendes Institute for Biodiversity Conservation (ICM-Bio) was created, bringing significant changes to current environmental education, especially in the area of environmental management. ICM-Bio originated from a split from IBAMA, which was divided into two institutions: IBAMA itself, responsible for assessing environmental impacts and activities linked to environmental licensing, and ICM-Bio, responsible for managing Conservation Units (Brasil, 2014).

In 2009, the 6th Brazilian Forum on Environmental Education was held at the Federal University of Rio de Janeiro (UFRJ), with the support of the Ministry of the Environment. Its central theme was the Treaty on Environmental Education, which was developed in that city in 1992 at the Global Forum (Brazil, 2014).

In 2012, CNE/CP Resolution No. 02/2012 was created, which regularised the National Curriculum Guidelines for Environmental Education, to be observed by education systems and their basic education and higher education institutions:

Art. 16 - The inclusion of knowledge concerning Environmental Education in the curricula of Basic Education and Higher Education can occur: I - through transversality, through themes related to the environment and socio-environmental sustainability; II - as content of the components already included in the curriculum; III - through the combination of transversality and treatment in curricular components (BRASIL, 2012).

2.1.2 . Institution of the National Environmental Education Policy

Due to the immeasurable importance of its implementation, Environmental Education has gained a place in Brazilian environmental legislation. The issue is dealt with most prominently in Law No. 9.795 of 27 April 1999, which establishes the National Environmental Education Policy (Brasil, 1999).

Art. Iº EE is understood to be the processes through which individuals and the community build social values, knowledge, skills, attitudes and competences aimed at conserving the environment, which is a common asset of the people, essential to a healthy quality of life and its sustainability (Brazil, 1999).

Art. 2º EE is an essential and permanent component of national education and must be present, in an articulated manner, at all levels and modalities of the formal and non-formal educational process (Brazil, 1999).

Art. 9º EE in school education is understood to be developed within the curricula of public and private educational institutions (Brazil, 1999).

Art. 10 EA will be developed as a continuous and permanent integrated educational practice at all levels and modalities of formal education (Brazil, 1999).

Art. 1 Iº The environmental dimension must be included in teacher training curricula at all levels and in all subjects (Brazil, 1999).

Art. 13 Non-formal environmental education is understood as educational actions and practices aimed at raising public awareness of environmental issues and at organising and participating in the defence of environmental quality. It should promote emancipatory education, i.e. education that stimulates people's autonomy (Freire, 2003), considering that:

> Every educational process is first and foremost a process of intervention in the lived reality in which educator and student, in a dialogical practice, construct knowledge about it, with the aim of transforming it (Brasil, 1999 p.9).

Each participant in the educational process brings their contributions to the collective, both from their experiences and from their formal (school), informal (learnt through social groups) and non-formal (learnt through collective actions) knowledge. This can make the process richer, closer to the reality of all those involved and, therefore, with a greater chance of continuity and success.

Interesting ways of realising the diversity and complementarity that we work on in EE, using just a few significant prepositions (Sauvé, 2001 p.317), are:

- Environmental education - informative, focussing on the acquisition of curricular knowledge, in which the environment becomes the object of learning. Although knowledge is important for a critical reading of reality and for finding concrete ways of acting on environmental problems, it is not enough on its own;

- Environmental education - experiential and naturalising, in which contact with nature or "walks" around the school are encouraged as contexts for environmental learning. With walks, nature watching, outdoor sports and ecotourism, the environment offers experiential experiences, becoming a learning environment;

- Education for the environment - constructivist; seeks to actively engage through socio-environmental intervention projects that prevent environmental problems. It often takes a critical view of the historical processes of building Western society, and the environment becomes the goal of learning;

- Education - starting from the environment - this considers, in addition to the others included, the knowledge of traditional and indigenous peoples that always starts from the environment, the interdependence of human societies, the economy and the environment, the simultaneity of impacts at local and global levels, a review of values, ethics, attitudes and individual and collective responsibilities, participation and cooperation, recognition of ethnic-racial differences and the diversity of living beings, respect for the territory with its carrying capacity, improving the environmental quality of life for present and future generations, the principle of uncertainty and precaution.

Twenty-seven years ago, actions in environmental education were not based on an adequate theoretical framework, as most of the publications were foreign translations, aggravated by the limited number of qualified professionals who could contribute to the debate (Sato, 1997).

Over the years, environmental education has shown significant growth, with increasingly complex formulations and diverse practices that represent a significant conceptual evolution (Reigota, 2006). Even through this concept, it faces didactic challenges materialised in various forms: universal models of intervention, which are inefficient because they reproduce a purely ecological vision and fail to take account of socio-environmental specificities, represent a major challenge. Everyday issues are sometimes left unaddressed, without being considered important in a given context.

Dialogue on environmental education is more than a scientific, political or epistemological issue; it is an ethical-anthropological issue of fighting for life (Freire, 2003). It requires changes in values, attitudes and actions in the way we interact with the environment and with life. It means taking on new ways of perceiving the world that go beyond the idea of conservation and involve a direct relationship with feeling, living, thinking and being, in order to protect life on the planet.

Humanity is rooted in a "homeland", the Earth, because, more than biological, humanity is inseparable from the biosphere (Morin, 2000). Education has to be a permanent process and, as such, it has to be an obligatory collective action, so that a collective existential stance is sought, as a proposal for activities aimed at building a new society.

2.1.2 Environmental Education in the National Curriculum Parameters

The National Curriculum Parameters (PCN) were developed from the context that the school is a place of information and training, with the learning of content having the purpose of contributing to the inclusion of students in everyday social issues, and that the school provides training aimed at developing skills.

The PCN are a quality reference for primary and secondary education in our country and are presented not as a curriculum, but as aids that support projects in the school, in the development of the curriculum programme. Its major innovations are the so-called transversal themes for primary education, which include the environment (Brasil, 1997).

After their contextualisation and officialisation, five themes were allocated to complement the content proposed by traditional subjects. These themes are directly related to the exercise of citizenship (Brazil, 1997) and should give "similar weight" to the contents of those subjects. We call all these themes "transversal themes" (Brasil, 1997). Transversality refers to the possibility of establishing in educational practice a relationship between learning theoretically systematised knowledge, permeating all grades, to help schools fulfil their role of educating students for citizenship (Souza, 2002).

Therefore, within the framework of the PCN, environmental education has the main function of contributing to the formation of conscious citizens who are able to decide and act in socio-environmental reality, committed to life and to the well-being of everyone in local, regional and global society. In order for this to happen, schools must develop proposals for working with attitudes and values, teaching skills and procedures that will be passed on to students as they go about their daily lives in their social environment, so that they can express their cultural and artistic values (Brasil, 1997).

2.1.3 Non-formal education

For Brandão (1981, p.6), education is everywhere; there can be social networks and structures for transferring knowledge from one generation to the next. The evolution of human culture has led human beings to pass on knowledge, creating social situations. These situations are necessary when we realise that "socialisation is responsible for the transmission of knowledge".

Libâneo (2005, p.27) states:

> In fact, the pedagogical power of various formal and non-formal educational agents is growing. Pedagogical actions take place not only in the family and at school, but also in the media, social movements and other organised human groups, and in non-school institutions. There is pedagogical intervention in television, radio, newspapers, comics, in the production of information material such as textbooks, encyclopaedias, tourist guides, maps, videos and also in the creation and development of games and toys.

Freitas and Martins (2005, p.2), analysing the emergence of education beyond school, state that:

The concept of non-formal education arose as a result of the new demands of social transformation, produced by scientific and technological advances, which have had significant economic, socio-cultural, political, demographic and consequently educational consequences and implications. From the mid-20th century onwards, there was a growing concern that the general public should have access to scientific and technological knowledge and that this knowledge should extend beyond school education.

The expansion of non-formal education began as a result of changes in the contemporary world, which ended up requiring pedagogical practices outside the school environment as a way of meeting this demand from students (Trilla, 2008).

Jacobucci (2008) understands that a non-formal education space may or may not be linked to an institution. In those that are institutionalised, there are precepts that establish how they work, as well as a group of people who work with the aim of achieving the proposed objective for the space, thus differentiating it from a non-institutionalised one; for example, a beach, a mangrove swamp or a square, which can also be useful for science education.

Since the 1990s in Brazil, these experiences have come to be known as non-formal education, taking place alongside formal education. As a result, schools have sought to review and rethink their ways of happening and existing at many points in history.

To characterise it, Marandino (2003) specifies some of the elements that are part of it: they carry out actions of a collective nature; those who join it act spontaneously; they have flexibility in their subjects as a fundamental feature, which can be used in different ways and on different occasions; they promote non-standardised events and do not set deadlines for learning something. For Vieira (2005), it generates learning about subjects offered by the school, but with different basic characteristics, such as its format, its conduct, its goal and, above all, its construction site: outside the school. Oliveira and Moura (2005) state that its purpose is to offer procedures and seek out spaces that allow knowledge to be obtained and improved.

According to Bento (2007), it is the set of actions carried out beyond the confines of the school, which take place in an orderly, regular manner, providing education and learning to defined groups.

Gohn (2010) considers that one of the great challenges of non-formal education has been to define it, characterising it for what it is. It is usually defined by negativity, by what it is not. It should be noted that non-formal education does not have the formal character of school processes, standardised by official higher institutions and certifying qualifications. It differs from formal education because the latter has national legislation that standardises specific criteria and procedures.

Still in the formal/non-formal binomial, there are authors such as Brennan (1997) who characterise non-formal education as a complement, an alternative space for school rebels and insubordinates.

For Trilla, non-formal education is the
> [...] a set of means and institutions that generate educational effects through intentional, methodical and differentiated processes, which have previously and explicitly defined pedagogical objectives, developed by agents whose educational role is institutionally or socially recognised, and which are not part of the graduated educational system or which, if they are part of it, do not constitute written and conventionally school forms (Trilla, 2008, p. 22).

Afonso (1992 and 2006) introduces the category non-school as a synonym for non-formal; however, he warns: "the justification of non-school education cannot be built against the school, nor serve any strategies for the destruction of political education systems". Cortella (2007) also adopts the same line and goes further: for him, non-formal education should be articulated with formal education and act in a complementary manner.

Non-formal education has its own field and purpose; its centrepiece should be "training for citizenship and the social emancipation of individuals". Non-formal education practices usually take place outside of school, in social organisations, social movements, community associations and human rights training programmes.

According to Sato and Santos (2001, p. 270):

> Most people don't understand the intimate relationship between human activities and the environment, due to ignorance or inadequate information. It is of fundamental importance to sensitise people and involve them in environmental problems, in order to seek effective solutions for environmental development and planning. The educational process can awaken the ethical and environmental concerns of human beings, changing values and attitudes, building the skills and mechanisms needed for sustainable development. For this to materialise, there needs to be a new reformulation of education, providing not only information about the physical and biological environments, but also about the socio-cultural environment and human development, thus seeking to improve the quality of life of individuals.

Libâneo (2005, p. 95) also emphasises the relationship between school and non-formal education when he states:

> Formal and non-formal education are constantly interpenetrating each other, since non-formal education cannot do without formal education, whether in schools or not, official or not, and formal education cannot be separated from non-formal education, since students are not just pupils, but participants in the various spheres of social life, at work, in the trade union, in politics and in culture. It is therefore always a question of interpenetration between school and non-school.

Non-formal education is more diffuse, less hierarchical and less bureaucratic. Its programmes, when formulated, can be of variable duration; the space category is just as important as the time category, because learning time is flexible, respecting biological, cultural and historical differences. Non-formal education is closely associated with the idea of culture (Gadotti, 2005).

Cortella (2007, p. 47) states that:

> As education is not synonymous with school, since school is part of education, everything that expands beyond school formalisation is educational territory to be operated in. Furthermore, if sharing in non-formal education aims to consolidate a society with fair and equitable coexistence, citizenship in peace is the horizon.

Vidal (2009, p. 28-29), when researching people with special needs in São Paulo, states that:

> The experiences made possible in the process of non-formal education can have a significant impact on this segment, not only as a possibility of complementing schooling, but also of awakening motivations and interests that help to promote their inclusion in formal education. In any case, such experiences are rich, even for those who have already completed compulsory schooling and who will have gained a greater chance of exercising their cultural rights.

Non-formal education has some of its objectives close to formal education, such as the formation of a full citizen, but it also has the possibility of developing some objectives that are specific to it, via the form and space where it develops its educational practice, reinforcing, collaborating with and complementing formal education, dedicating itself to tasks that have nothing to do with it.

CHAPTER 3

OBJECTIVES

3.1. General Objective

- To evaluate the application of the interventions with 7º and 8º year-old primary school students from a municipal school, using the park's trails as a way of raising awareness among these students, so that they could identify the environmental issues of the Conservation Unit.

3.2. Specific objectives

- Approach interpretive trails as a means of teaching science in non-formal spaces;

- Understanding that well-planned pedagogical actions (group dynamics) create favourable environments that encourage students to build on their prior knowledge;

- Relate the knowledge visualised along the way to the knowledge acquired in formal education with the Science, Geography and History teachers (multidisciplinary);

- Reflect on the importance of teaching resources for science lessons, specifically environmental education, in non-formal spaces.

CHAPTER 4

INTERPRETIVE TRAILS

The teacher with a guide/teacher's vision is where the question of interpretive trails as a tool for environmental education comes in. Long before trails had an educational function, their main function was to fulfil the need to travel, like roads made centuries ago to connect one town to another; but over the years, there has been a change in values in relation to trails. Instead of travelling, trails appear as a new way of connecting with nature. The various concepts of interpretive trails (Belart, 1978) consider that ["walking, hiking, climbing, touring, away from the hustle and bustle, crowds, noise and vehicular traffic is nowadays one of most people's favourite pastimes. It's the cheapest, healthiest form of recreation and offers the greatest opportunities for observation, research, tranquillity and daydreaming"]. Trail is a word from the Latin "sequor", meaning path, course, direction. Over the years, humankind has opened up and used these paths to meet its needs, mainly for travelling. Nowadays, however, trails are being used as a way of getting in touch with nature, living in harmony and feeling well. According to Vasconcellos and Ota (2000), "a trail is a path through geographical, historical and cultural space". The way in which this translation of the meaning of trail is made, the interpretive approach, is what differentiates interpretation from the simple communication of information.

The trail is considered interpretive when its resources are translated for visitors, based on predefined themes, through specialised guides, leaflets or panels. In practical terms, interpretive trails are intended to stimulate groups of authors into a new field of perception, with the aim of getting them to observe, question, experience, feel and discover the various senses and meanings related to the theme (Vasconcellos, 1998).

Within the work with interpretive trails, Freeman Tildem is considered the founder of interpretation, with the Doce Matas Project, 2002. For him, "Interpretation is an educational activity that aspires to reveal the meanings and relationships that exist in the environment, by means of original objects, through first-hand experiments and illustrative means, rather than simply communicating literal information". Trails have been one of the most widely used means of environmental interpretation, both in natural and built environments.

Interpretive trails exist not only to communicate facts, dates and concepts, but also to share experiences that lead visitors, whether students, teachers or tourists, to appreciate, understand, raise awareness, co-operate in the conservation of a natural resource and also to educate (Menghini, 2005). With different visions of the environment, environmental interpretation continues to be a translation of the language of the environment into the common language of people, enabling them to perceive the world in a way they have never done before.

One of the pioneering park rangers who began this vision of guiding people to places and making them aware of their surroundings was Enos Miles; he worked in the world's first registered park as a Conservation Unit, Yellowstone National Park in the USA. In 1922, Miles said that an interpreter is a naturalist who knows how to guide others to the secrets of nature. As such, the use of these trails becomes a pedagogical attraction for carrying out educational practices, with defined objectives for building the concepts, values and worldviews of the different audiences that visit these spaces (Menghini, 2005).

According to Mello (2006), an important method of environmental education is to transform theoretical lessons taught in classes into extramural activities, using the resources present in "Interpretive Trails". These are often used in projects as a means of environmental interpretation, aiming not only to transmit knowledge, but also to provide activities that analyse the meanings of the events observed in the environment, as well as its characteristics (Zanin, 2006).

The trails that exist for educational purposes are usually short and can be defined as a nature trail that promotes closer contact between human beings and nature. They are an important pedagogical tool that enables knowledge of flora, fauna, geology, history, geography, biological processes, ecological relationships, the environment and its protection, as well as the development of attitudes and values in individuals. For students visiting natural areas, trails translate factors that go beyond appearances, such as natural laws, interactions, functioning, history or facts that, although obvious, are not commonly perceived by those walking along them (Dias and Zanin, 2004).

An important point for individuals to gain knowledge is to develop awareness. When discussing perception, we can see the essence of "place", which corresponds to the concrete manifestation of "space", where the set of uses and habits projects the image of this space as a daily life, often preventing its perception. Overcoming the vision of everyday life is a prerequisite for environmental awareness, because we understand that knowledge of the whole is capable of inducing new knowledge, promoting changes in behaviour (Tuan, 1980).

We can't talk about landscape without dealing with perception, because understanding the space perceived through visual perception is of fundamental importance. The individual is encompassed by the landscape and cannot say that it is around them, offering them limitations as they observe it, because there are visible and non-visible portions that make up the same landscape. "Landscapes are not separate from human experience. It is man who experiences landscapes, attributing meanings and values to them" (Machado, 1998). Humans, as just another species on the planet, have no right to destroy other forms of life. Humans, as "the only beings capable of understanding the greatness of the phenomenon of organic evolution, have the inalienable ethical duty to allow it to continue its course and for biological diversity to remain exuberant" (Câmara, 2001).

Currently, the problems inherent in environmental issues are intensifying and, at the same time, initiatives are being taken by various sectors of society to develop activities, projects and the like in the quest to educate communities, raising their awareness of environmental issues and mobilising them to change harmful attitudes and adopt attitudes that are beneficial to environmental balance (Milano, 2001).

The search for an understanding of the balance of the environment is becoming one of the strongest trends today, as society is already feeling the effects of the serious degradation of natural resources and landscapes. Therefore, educational activities in environments with significant ecological potential and great biodiversity can become an important tool for the preservation and conservation of these spaces (Jesus and Ribeiro, 2006).

The establishment of the first national park in 1872 was an important step towards the creation of conservation units in other countries. Brazil's first national park, Itatiaia, was founded in 1937. It is estimated that there are 1930 Conservation Units (CU), each classified according to its characteristics and the objectives to be achieved - National Register of Conservation Units (Brazil, 2014). They can be used for the sustainable exploitation of natural resources, total preservation of the ecosystem, research and visits to promote environmental education.

The National System of Conservation Units (SNUC) was established by Law No. 9.985, of 18 July 2000 (Brasil, 2014), and consolidated the regulatory framework for Conservation Units in the country, bringing together the existing instruments and rules used until then. This system, complemented by Decree No.⁰ 4.940 of 22 August 2002 (BRASIL, 2014), established the criteria and standards for the creation, implementation and management of Conservation Units.

They are called National Parks, Biological Reserves, Ecological Reserves, Ecological Stations, Environmental Protection Areas, Areas of Relevant Ecological Interest, National Forests, Extractive Reserves, Wildlife Refuges, Fauna Reserves, Sustainable Development Reserves and Private Natural Heritage Reserves (Brazil, 2014).

The area selected for this study is characterised as a Municipal Nature Park, according to Chapter III, Article 11 (Brasil, 2014), and should enable scientific research and the development of environmental education activities, environmental interpretation, recreation in contact with nature and ecological tourism. It is made up of three ecosystems rich in biodiversity and of great environmental importance in the metropolitan region, especially the third, which is coastal and plays an important role in the reproduction of marine fauna. It is a region subject to periodic flooding, where swamps and mangroves predominate. Despite the adverse conditions, there is urban occupation - many areas have been built up with housing.

Krasilchik (2000) made a brilliant reflection on the role of environmental education: to function as an integrating element, so that the community becomes aware of the phenomenon of development and its environmental implications. "To this end, it should serve not only to impart knowledge, but also to develop skills and attitudes that enable man to act effectively in the process of maintaining environmental balance, so as to guarantee a quality of life that is consistent with his needs and aspirations." (...) In order for environmental education to fully achieve its objectives, certain aspects must be considered: providing students with a solid base of knowledge; however, knowledge alone is insufficient.

CHAPTER 5

DEFINING SENSITISATION

The term sensitisation comes from sensibility, which comes from the Latin *sensibihtas*. According to Bueno (2014) and Ávila, (1972), it means the act or effect of sensitising, of making sensitive or susceptible, and sensitiser means "to make sensitive, moving or touching". Thus, we can understand that sensitisation can be produced, and the important factors for promoting sensitisation are: using activities that can inspire conscious and attuned attitudes; individuals working in groups, who have experience and points of view that share the same purpose and development process. Awareness-raising is the first general basic principle of environmental education (Smith, 1995), and the process of alertness is the first step towards achieving systemic thinking about the educational environmental dimension.

Londôno and Guerrero (1999 p. 34) comment that:

> [...] sensitisation has an epistemological meaning, that is, it studies the degree of certainty of scientific knowledge in its various forms. It's the first action of human knowledge: it's getting closer to the object based on the knowledge and action that one must have of reality. It means getting close to nature, feeling it, experiencing it and studying it.

Regarding the use of some tools as strategic teaching actions in the practice of environmental education, based on awareness-raising, and according to Berger (1999, p. 54), we can highlight:

> Visits to museums, scientific wild animal breeding sites and walks on ecological trails, where the trails are usually interpretive; they have routes in which there are specific points for interpretation with the help of signs, arrows and other indicators, or spontaneous interpretation can be used, in which the trail guides stimulate the students' curiosity as events, places and facts occur, made through direct observation of the environment, and the drawings that become effective tools for indicating the themes that most stimulate the observer's environmental awareness.

According to Tristão (2002), teachers, as mediators, re-elaborate or give new meanings to the daily information they receive. In this sense, it is not enough that students have all the information, but that they learn to interpret it. With regard to students' interaction with the environment, and as a way of initiating an educational process (sensitisation), we find ourselves thinking along the lines of Tiriba (2004), when he states that no one will be able to love and preserve what they don't know; a nature with which they don't live interactively. To this end, it is necessary to provide a real approach through the students' everyday relationships with water, rubbish and environmental degradation, taking elements present in the scenarios of the teaching activities being provided.

In this way, the practice of raising awareness is one of the ways to carefully implement and adapt the broad and contextualised knowledge to be passed on, as Folena and Anjos (2005 p.34) point out:

> We cannot turn environmental education activities into a purely informative practice involving content that is disconnected from a meaningful context. Such a practice would not consist of learning, but rather the development of a negative and uncompromising expectation of the environment.

This means, according to Pinheiro (2005), that the environmental problems resulting from any issue are prioritised, but it is important to stress that others are by no means unimportant either, especially those related to the central theme, which should not be neglected from the discussion agenda, such as the issue of solid waste management (rubbish), water contamination, riparian forest degradation, and others.

The various educational practices and actions that have been developed to raise public awareness about environmental issues, in which schools, universities, companies and other institutions participate and partner, are taking on a huge dimension in Brazil. It is a strong instrument for changing the behaviour of individuals who, once they have been made aware, are integrated into their communities, schools and even their work environment, in defence of their causes, which do not degrade the environment in which they live. EE activities, as instruments to be used to achieve the objectives, should provide participants with opportunities to develop awareness of their environmental problems, as well as reflect on solutions.

In order to raise awareness, as a continuous process of environmental education, it is of the utmost importance to define the issue to be studied and the activities to be developed in projects and programmes. These activities should be integrated with the systematised knowledge and reality of the individuals involved, thus leading to a process of awareness, commitment and environmental awareness, as well as learning about the topic to be developed, thus enabling the development of competence, analysis, decision-making, planning and research, in other words, all the requirements that individuals need to fully exercise citizenship.

5.1. Environmental Interpretation

Many people are familiar with interpreting texts, and yet this word can have a range of meanings for different individuals, depending on their education, training or work experience as an interpreter. To date, many authors have released their judgements of what environmental interpretation is; therefore, there is no single definition; on the contrary, there are many, each with different approaches.

Aldridge (1973) understands environmental interpretation as the art of explaining man's place in his environment, in order to increase visitors' awareness of the importance of these interactions and awaken in them a desire to contribute to environmental conservation.

Interpretation is an educational activity that aims to reveal meanings and relationships through the use of original objects, direct contact with resources or illustrative media, and is not limited to providing information about facts. Tilden (1977).

According to Morales (1992), rather than trying to explain, reveal or inform, interpretation satisfies curiosity and is not a way of educating the public, it must be sufficiently suggestive to encourage the individual to change their attitude or adopt a certain stance. Environmental interpretation should be recreational, so there is no rejection of the interpretive proposal.

Environmental interpretation argues that it involves translating the technical language of a natural science or related field into easy-to-understand terms and ideas for people who are not scientists; and doing so is fun and interesting for them. Ham (1992).

Environmental interpretation is not just information; it is communication that takes place through direct contact with other media; it is the use of techniques designed to arouse interest, a change in attitude, the visitor's knowledge and appreciation of the resource being interpreted by establishing direct contact with the environment. However, all these definitions cannot, in essence, take environmental performance out of a simple and clear way of translating nature. Ham (1992).

5.1.1. Types of Interpretive Objectives

Management objective - To be able to facilitate the fulfilment of management goals. Firstly, interpretation can alert visitors to the proper use of recreational resources, helping to reinforce the idea that parks are special areas that require special behaviour. This objective is especially supported by the previous ones. It is very important because interpretation can be used to minimise human impact on resources. Sharpe (1988).

Learning objective - These are things that the visitor can point to and identify. Most visitors will be able to discover the biodegradation process of three common everyday elements (soft drink can, aluminium and paper). The essence is to give pleasure and education to the visitor. Interpretation should help make life a rich and enjoyable experience. It can increase the visitor's enjoyment, so that there is a better understanding of the place and increase the pleasure derived from the visit. Moore et al (1989).

Behavioural objective - This is an amortising objective, a true project purpose; interpretation creates conservation awareness in the visitor. Moore et al (1989).

Any intention to provide interpretation implies a very clear goal, common to other management and administration activities in an area: the conservation of its natural resources. This conservation can only be achieved through the respect and citizen participation that interpretation aims to provide. Aldridge (1973).

5.1.2. . Environmental Education Activities

If we want an "attitude of reverence towards life", first of all we need to develop perception, which in turn can be transformed into love and empathy. As we begin to feel what unites us and what is common to all living beings around us, our actions become more harmonious and spontaneous, so that we naturally become more aware of the needs and well-being of all living beings. Comell (1997).

Environmental education activities on interpretive trails can be practical tools that help raise students' awareness and experience of the complexity of environmental issues, creating moments for them to reflect on their experiences in a creative, fun and stimulating way. Preferably, these activities are developed to introduce a theme, creating the conditions for receptivity and interest in the content worked on later, and should be used within a broader educational context that involves the complexity of environmental issues. Menghini (2005).

We agree with these authors: raising awareness of care and respect for the earth, life, human rights, economic justice and a culture of peace are also part of the educational process, which is not just about cognitive and methodological aspects. This dimension is emphasised in documents such as the Earth Charter. Maurice and Mikhail (1997).

CHAPTER 6

METHODOLOGY

The work was carried out between 20 November 2013 and 24 October 2014. We chose to carry out the research, which is classified as both qualitative and descriptive. Carried out in a public school in the municipality of Magé, in the Ipiranga neighbourhood, it is part of the research project "Basic Education in Non-Formal Spaces in the Baixada Fluminense", which was supported by FAPERJ.

Educational research carried out by primary school teachers has increased dramatically in recent years. With the possibility of experiencing diverse situations in everyday school life, it offers teachers a very rich and varied range of topics.

The municipality of Magé is of great importance from an environmental point of view, although the municipal government has defined the CU, in the conception of the bodies linked to environmental issues, from a viewpoint that remains preservationist and neglects the socio-economic dimension of the environment.

6.1. Research characterisation

Qualitative research

It presents five basic characteristics that serve as a reference: "1. the natural environment as a source of data, and the researcher as the main instrument; 2. the data collected is predominantly descriptive; 3. the concern with the process is much greater than with the product; the meaning that people give to things and to their lives are the focus of the researcher's special attention; 4. the researcher must be attentive to capturing the participant's perspective; 5. data analysis tends to follow an inductive process." Ludke and André (2014). Furthermore, "qualitative research involves obtaining descriptive data through the researcher's direct contact with the situation studied; it emphasises the process rather than the product and is concerned with portraying the perspective of the participants." Bogdan and Bilklen (2001).

Descriptive Research

Cervo and Bervian (2002) characterise it as research that seeks to understand the various situations and relationships that occur in social, political and economic life, aspects of human behaviour, both individual and group, and more complex communities without manipulating variables. Gil (1995) states that descriptive research aims to "describe" the characteristics of the population, facts and phenomena.

The research is characterised as descriptive, since no experiments will be carried out with the manipulation of variables. The purpose of the research is to describe the data found, which, based on Lakatos and Marconi (1992), also covers four aspects: describing, recording, analysing and interpreting current phenomena, with the aim of understanding how they work in the present.

Booth (2000) states that the primary objective of descriptive research is to describe the characteristics of a given population or phenomenon, or to establish relationships between variables. In descriptive research, the facts are observed, recorded, analysed, classified and interpreted, without the researcher interfering with them, i.e. the phenomena of the physical and human world are studied, but not manipulated by the researcher, a characteristic that this research takes into account.

6.2. The Barão de Mauá Municipal Nature Park

The Barão de Mauá Municipal Natural Park, where the research took place, is a protected area. It was created by municipal decree 2.795/201; it is located in the Ipiranga neighbourhood, in the municipality of Magé, in the Baixada Fluminense (BF), which is part of the Metropolitan Region of Rio de Janeiro. It has a total area of approximately 116.80 hectares, occupied entirely by the mangrove ecosystem, with the mouth of the Estrela River, the Estrela Environmental Protection Area (APA), the water mirror of Guanabara Bay and the dry land and urban areas surrounding the mangrove as its boundaries (Fig.l). Over the years, the park has suffered various impacts due to the removal of native vegetation, crab hunting, the dumping of domestic sewage and the amount of rubbish brought in by the tides. The oil spill in Guanabara Bay in 2000 caused negative environmental impacts in the park area. In 2001, the NGO OndAzul began a project to reforest the mangroves in the UC areas. In just over 10 years, more than 25 hectares had already been reforested. This project led to the transformation of the area into a UC, which is under the administration of Magé's Municipal Secretary for the Environment. In Brazil, mangroves are protected by federal legislation, due to their importance in the marine environment, as they are fundamental for the breeding and growth of the young of various animals, as well as being used as a migratory route for birds and for feeding fish. The main law that protects mangroves is the Forest Code (Law No. 12.651/2012), which, in item VII of Art. 4° , defines mangroves, throughout their entire length, as a Permanent Preservation Area (APP).

FIG: 1 Barão de Mauá Municipal Natural Park and Prof.ª Hilda da Silva Coelho Municipal School Source: Google Earth, 2015

6.3. Developing Research

Authorisations

In November 2013, the first contact was made with the Secretary of the Environment of Magé - RJ. In the same month, contact was made with the guide responsible for the UC. Contact was made with the management of the Municipal Elementary School "Professora Hilda de Souza Coelho", located in the Ipiranga neighbourhood of Magé, so that the idea of the research could be explained and also to find out if the school was interested in taking part. The initial idea was explained, emphasising that the participation of the students would be fundamental for the work to be carried out. The headmaster responded positively and was keen to contribute. The 7° and 8° year olds were introduced to him, and he made his intention clear at that point, i.e. to understand what they understood about certain issues related to the environment. The aim of the research was briefly explained (Tab.l).

Date	Activities

20/11/2013	Visitation by the researcher to the UC to contact the guide.
10/03/2014	Visits by 7th graders° to get to know the CU.
14/04/2014	Visit by 8th graders° to get to know the CU.

Tab. 1 The first contacts.
Source: Vieira

Research Subjects

The subjects of the research were the students living in the area surrounding the UC, comprising 33 students from the 8th grade° and 27 students from the 7th grade° of primary school II at the municipal school in the Ipiranga neighbourhood - Magé.

Data collection

Data collection took place between February and October 2014, and consisted of four educational activities in the UC. The research instruments were:

- Use of a questionnaire to evaluate the students' educational activities after visiting the non-formal educational space;

- Group dynamics on the Interpretive Trails;

- Observation of educational activities in non-formal spaces.

Questionnaires

Lakatos and Marconi (1992) point out some advantages and disadvantages of using questionnaires. The advantages are: less expensive activity; saves time and obtains a large amount of data; reaches a greater number of informants; has a lower risk of distortion; the impersonal nature of the instrument, facilitating evaluation. The disadvantages are: unanswered questions; influence of one question on another; impossibility of support for the questions.

Application of the instruments

The student questionnaire (after the visit) was administered on the same day as the field trip, so as not to jeopardise the teacher's lessons, who would have to return to the school to teach the classes that were not part of the project (Tab. 2).

Date	Class	Time	Students
15/05/2014	7° year	9am-12pm	27
10/10/2014	8° year	9am-12pm	33
Total			60

Tab. 2: Application of the survey instruments.
Source: Vieira.

6.4. Ethical Research Procedures

In accordance with Opinion No. 658.905 of 22 May 2014, Unigranrio's research ethics committee, in compliance with CNS/MS Resolution No. 466/12, approved the project at a meeting held on 22 May 2014.

According to Sandín Esteban (2010), in qualitative research, the criteria involved in its credibility and validity require ethical implications to be taken into account. In this type of research, there is a great deal of interaction between the researcher and the subjects involved, making it necessary for certain procedures to be adopted in order to guarantee quality.

With regard to the issues addressed, certain precautions were taken. Regarding the fieldwork, there was a previous conversation with the school headmaster and the science teacher. The principal authorised the research. With regard to the students, during the first observation in each class, the subject teacher himself explained the presence of the researcher and the purpose of being there in that school environment, so there were no objections to this issue.

With regard to the photographic recordings with the students, the parents were sent the Informed Consent Form (ICF), as required in the biomedical field by the Code of Ethics.

Authorisation to Conduct Research in the CU

According to the letter of consent UC 001/2014- Parque Natural Municipal Barão de Mauá, designated by the Magé Secretary of the Environment - the managing body of the Parque Natural Municipal Barão de Mauá Conservation Unit, there is nothing to oppose in carrying out the research project "Parque Natural Municipal de Magé: Recurso Pedagógico para Sensibilização Ambiental" (Magé Municipal Natural Park: Pedagogical Resource for Environmental Awareness).

6.5. Activities Carried Out in the Non-Formal Space

6.5.1. Science students' learning in non-formal spaces

The activities were observed in a non-formal space, a questionnaire was administered and the students were monitored. The results of this research will be described below. Four activities were monitored during the course of the research:

6.2.1. Activity 1 - Mini-lecture
The first field activity took place on 15 May 2014, from 9am to 12pm, at the UC, with the Year 7° class (Fig. 4), reaching a total of 27 students, and on 10 October 2014, from 9am to 12pm, with the Year 8° class (Fig. 5), reaching a total of 33 students. It's worth noting that not all the students in the classes took part - only those whose parents authorised it. The visit was accompanied by the science teacher of the classes taking part in the research, the UC guide and the researcher. The students were gathered at the entrance to the park for a mini-lecture, in which they were explained what a Conservation Unit is, what its attributes are, the historical and cultural importance of the park, illustrated with panels, posters and maps. Afterwards, the park guide clarified the students' doubts. The activity lasted around 50 minutes.

Fig. 2: Year 7 students° at the entrance to the UC for the mini-lecture. **Photo:** Vieira (2014).

Fig. 3: Year 8 students° at the entrance to the UC for the mini-lecture.
Photo: Vieira (2014).

6.5.3. Activity 2 - Walking the Trails

The second activity began, a walk along the interpretive trails. The walk took around 50 minutes.

During the walk, the students stopped at different points to get to know some species of plants and animals, differentiating and clarifying their biological and ecological importance (Figs. 6 and 7).

A major concern is to pass on this technical information in such a way that the students understand what is being shown. Only native species are found on this trail, where their history, origin and development are explained.

6.5.4. Activity 3 - Environmental Education

A very interesting aspect of the walk along the trails is that you can see a contrast between the forest and part of Guanabara Bay; it opened up the opportunity to discuss different situations regarding the conservation of the riparian forest, the degradation that occurs from rubbish being deposited on the river bank. This open space made it possible to discuss the correct way to treat the environment in relation to rubbish; the students shared their experiences about the correct destination of the rubbish they produce. The park guide discusses the importance of environmentally friendly practices and the need to review concepts in order to find solutions for a sustainable future (Figs. 6 and 7).

6.5.5   Activity 4 - Questionnaire

At the end of the trail, the students gathered again at the park entrance so they could answer a questionnaire with questions related to visiting the park (Figs. 8 and 9). This activity, as well as bringing them closer together and sharing information, was also a way of encouraging them to investigate the topics covered. The questions included: Is this your first time hiking in the park? Did you enjoy the trail? Why? Did you like the guide's explanations? Would you like to visit the park again? What did you find most interesting along the trail?
This activity lasted approximately 20 minutes.

Fig. 4: Year 7 students walking the interpretive trails. **Photo:** Vieira (2014).

Fig. 5: Year 8 students° walking the interpretive trails.
Photo: Vieira (2014).

Fig. 6: Year 7 students° - Environmental Education Photo: Vieira (2014).

Fig. 7: Year 8 students° - Environmental Education **Photo:** Vieira (2014).

Fig. 8: Year 7 students° gathered at the entrance to the UC to answer the questionnaire. **Photo:** Vieira (2014)

Fig. 9: 8th grade students gathered again at the UC entrance to answer the questionnaire. **Photo:** Vieira (2014).

CHAPTER 7

RESULTS AND DISCUSSION

The following analysis refers to the results obtained by applying the questionnaire after the students' visit. The main aim was to gain a deeper understanding of the group's conception of the park and to point out the existence of another, less individualistic way of thinking about the environment. Based on the answers given, analytical graphs were developed that give us an insight into environmental issues and environmental education in non-formal spaces, which is increasingly being used with the aim of promoting a new environment in which to live, and not just within the scope of formal education. It is necessary to work in non-formal environments that can serve as places for differentiated learning.

Is this the first time you've hiked in the CU you've visited?

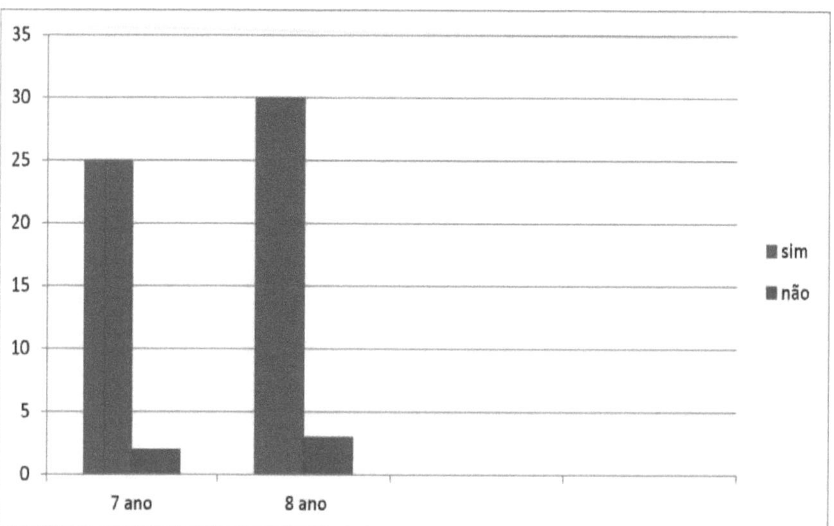

Graph 1 Analysis of the answer to question 1 of the questionnaire (visitation to the PA). **Source:** Vieira (2014).
Analysing the graph, we can see that 55 students said it was the first time they had visited the CU, even though they live in the park's surroundings - which demonstrates the importance of working on environmental education.

What do you expect to find on the park trail?

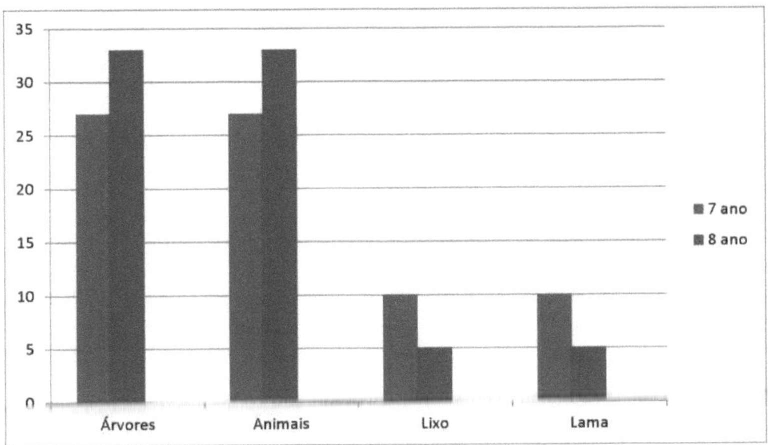

Graph: 2 Analysis of the answer to question number 2 of the questionnaire (getting to know the ecosystem of the CU).
Source: Vieira (2014).

Despite the lack of knowledge about the UC, the graph shows that the majority of students are interested in learning about this ecosystem by visiting it.

7.1. Attitude analyses

According to Ferreira (2005), our attitudes are related to the interests and values we construct in our interaction with the world. The aim of this analysis is to identify students' attitudes towards the preservation of Barão de Mauá Park from an ecological point of view. What environmental problems exist in the park? Who is responsible for the degradation of the park? What suggestions would you make to alleviate the park's degradation? Figure 14 shows the main environmental problems in the park cited by the students. According to the answers, the main problem is rubbish, followed by water pollution.

What environmental problems are there in the park?

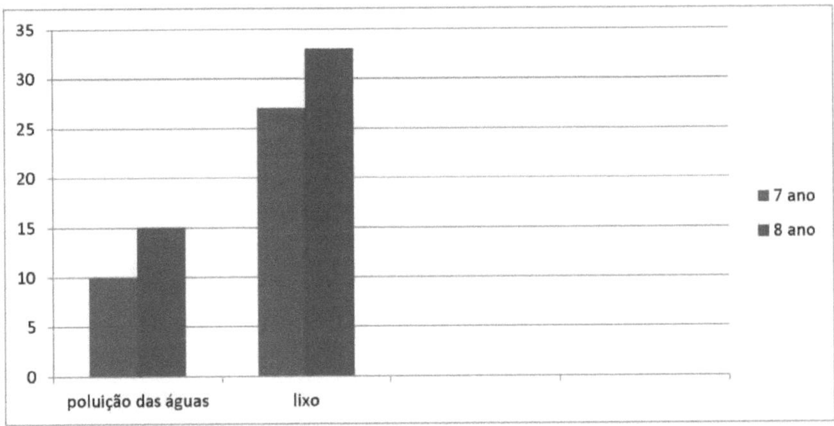

Graph: 3 Analysis of the answers to question number 3 of the questionnaire (environmental problems of the CU).
Source: Vieira (2014).

Fig: 10: Rubbish in the park from Guanabara Bay.
Source: Vieira (2014).

Fig: 11: Degradation of Guanabara Bay - a consequence of excessive rubbish.
Source: Vieira (2014).

Fig: 12: Barão de Mauá Park trail - evidence of lack of maintenance. Source: Vieira (2014).

When asked who was responsible for the park's degradation, the categories most expressed by the majority were the government and then people. For 25 of the students who took part in the survey, it was the municipal government that was responsible for the degradation, since it doesn't take any action to protect and conserve the park; for 22 students, it was the people who were responsible; 13 students said they didn't know who was responsible.

Who is responsible for the degradation of the park?

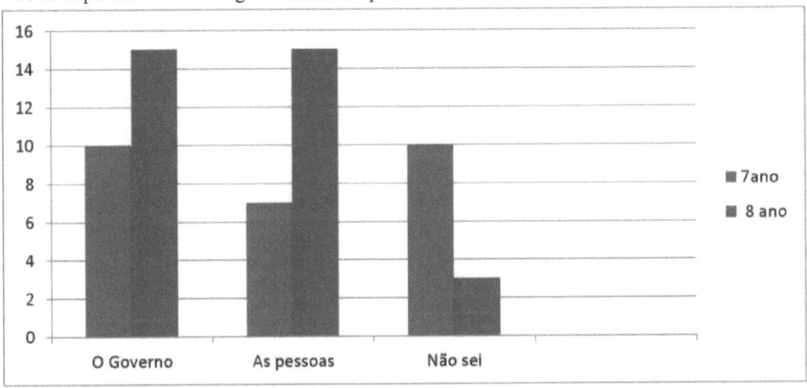

Graph 4: Analysis of the answers to question 4 (Who is responsible for the degradation of the Conservation Area?). **Source:** Vieira (2014).

What suggestions would you make to reduce the degradation of the park?

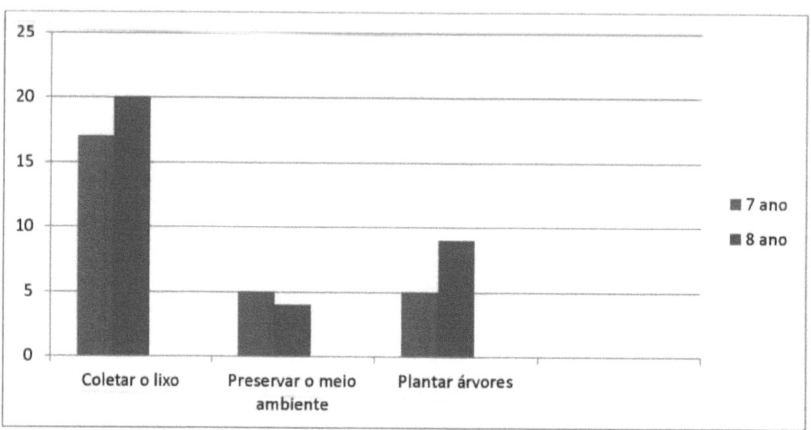

Graph: 5: Analysis of the answers to question 5 of the questionnaire (decrease in degradation of the CU).
Source: Vieira (2014)

In answers 3, 4 and 5, the students showed a partial perception of the exhaustibility of the environment. Guimarães (2003) states that society's recognition of the environmental crisis creates a consensus on possible solutions to this crisis. However, this consensual view is based on a conservative conception of society and the school itself, and is carried through to environmental education. It's important to realise here how relevant the issue of rubbish is for the students. The degradation of the park as a result of excessive rubbish being dumped in Guanabara Bay is the environmental problem that attracts the most attention. The park's situation, as well as that of the entire neighbourhood, is peculiar. Thus, the students' answers to these questions reflect the reality of environmental injustice in which the residents-students living around the park live.

The environmental education activities practised at school are not very effective in raising awareness of the social relationships that frame the process of environmental degradation. The answers to questions 4 and 5 suggest that the students assume a position of individual guilt for the crisis and believe that preservation actions, such as rubbish collection, alone can alleviate the degradation that is occurring within the mangrove biome. Another aspect that can also be taken into account is the unequal spatial relations that have been historically constructed in the region. The population with the lowest purchasing power has often, or almost always, had to build their homes in environmentally risky places.

Due to the lack of administration, which is visible in environmental degradation such as the accumulation of rubbish, the removal of firewood, bridges over the trails (without the minimum conditions for use), hunting and fishing, probably caused by members of the community surrounding the park, we suggest an emergency EE policy in the region's schools and the community, in order to raise awareness and thus generate environmental benefits and a better quality of life.

7.2. Analysing the Park's Environmental Value

In order to value the environment, students were asked about: 6- whether or not they liked the trails; 7- what activities could be developed; 8- whether they would like to visit the park again. The concepts of topophilia and topophobia (qualitative and evaluative) developed by Tuan (1980) mean, respectively, "affection or aversion to the environment" on the part of human beings. The results show that the topophilic aspect was identified in 100 per cent of the responses from the total number of students (60). This shows that, after the activity, the "research subjects" developed a strong affective involvement with the place, as evidenced by the "illustrative phrases" transcribed below:

"Yes, because you don't see what I saw every day."

"Because we can learn lots of new things."

"Yes, because it's rebuilding Maua, many people survive by collecting crabs and it helps to improve the environment."

"Yes, the vegetation, walking outdoors, meeting interesting animals. Among other things."

"Yes, the history of the park is very interesting."

"Yes, because it was interesting to study the mangrove and to study what we don't study at school."

"*Yes, because we know a little bit about each thing and we can have a person to explain* it *to us.*"

"Yes, I liked it because I feel very good about the environment."

"Because as well as explaining, the teacher teaches how to look after the mangrove".

"Yes. Because the whole thing was interesting, I've been on everything and it's great to be on everything."

"Yes. Because it's well organised."

"Yes. Because it was interesting to study the mangrove and to study what we don't study in schools."

"Yes, because she teaches me not to pollute, how to plant and how to respect animals."

According to Ferreira (2005), anthropogenic activities considered to be degrading the landscape are activities that can be resolved with greater involvement from society and management by public authorities. According to Sato (1997), the negative effect that human beings have on the environment puts life and the surrounding nature at risk.

From the answers given, we can therefore see the need to work with the groups surveyed on the beauty and importance of the place where they live, on the biodiversity that exists in the park, right next to their homes and their school, showing that there is not just degradation around them, but a rich environment waiting to be valued.

Did you enjoy the park's trails?

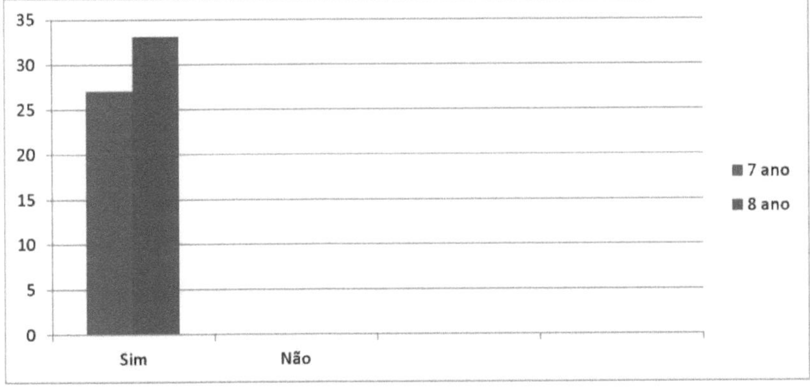

Graph: 6: Analysis of the answer to question 6 of the questionnaire (Did you like the course?).
Source: Vieira (2014)

When asked what activities could be carried out in the park, 60 per cent of the students said it could be environmental education. With this data, it can be inferred that, despite the existing degradation due to the accumulation of rubbish, the importance of environmental education is recognised by the majority of the students interviewed in the survey, as it is an activity that, although it needs planning, already exists on site.

What activities can be carried out in the park?

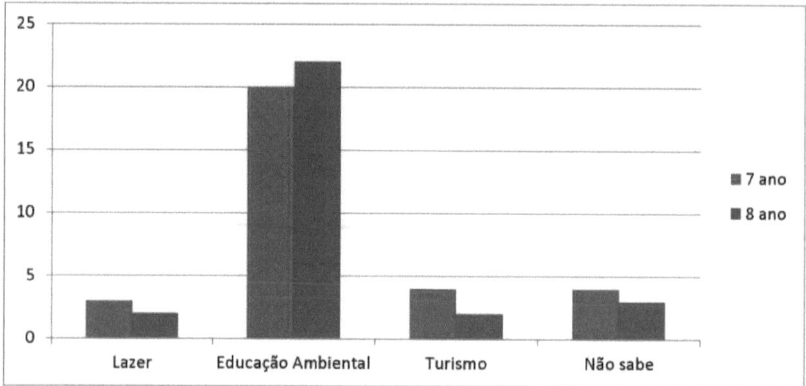

Graph: 7: Analysis of the answers to question 7 of the questionnaire (activities to be developed in the CU).
Source: Vieira (2014).

Would you like to visit the park again?

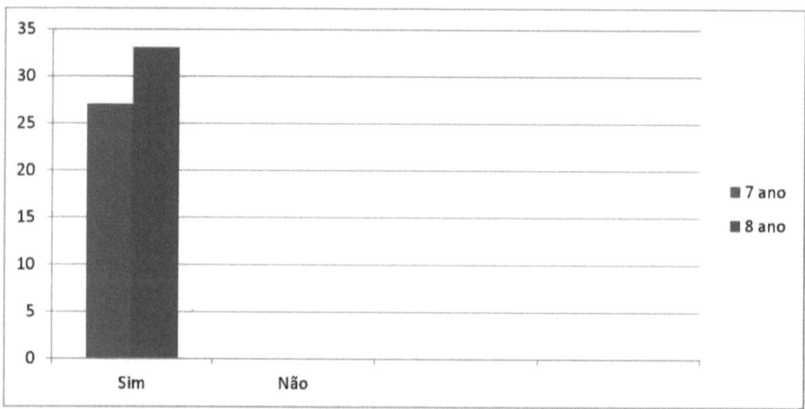

Graph: 8: Analysis of the answer to question 8 (Would you like to visit the CU again?). **Source:** Vieira (2014).

CHAPTER 8

FINAL CONSIDERATIONS

During the course of this research, some difficulties were outlined. The idea of working on "environmental awareness" with a group of primary school students who live around the UC in the Ipiranga neighbourhood, Magé-Rj, with the aim of understanding environmental issues within the context of these teenagers' lives, led to the question of whether they generated concerns and worries; what kind of action should be taken in relation to the future of the park; whether there was any form of involvement or participation in the search for solutions to environmental problems.

One of the difficulties encountered in the research was realising that the students were unaware of the existence of UC. But this difficulty was of such importance that it led us to discover that this whole process would be much more enchanting and enriching than immediate answers.

Non formal education takes place outside the classroom and its diversity affects each student differently. Understanding that one student is more of a motorist than a theoretician, they need to see phenomena in practice so that they can understand, interpret and sensitise them. Other students who are introspective or shy may develop their communication skills, as they will have to ask questions orally. The development of environmental projects in this space favours what greatly adds value to human learning: thinking about oneself in today's world, on a planet that is constantly calling for more ecological attitudes.

Through this research, we can see that awareness is fundamental to understanding the interrelationships we have with others, with society and, consequently, with the environment in which we live. It influences our expectations and attitudes towards environmental issues.

In this way, confirming some aspects of the theory that the purpose of "interpretive trails" is to develop a sense of awareness in visitors, it can be seen how the group surveyed perceived the environment and environmental problems in the areas they visited. As Vasconcellos (1997) has shown, an interpretive trail is a means to an end. A trail becomes interpretive when its relevant points and resources are shown to the people who will be using it, through signs, guides, teachers and others.

Another important point was the valuation of the environment, how it was seen by the students, whether this interaction with the environment generated any action or not, whether there was a real appropriation, an affectivity, a "topophilia", to use the term used by Tuan (1980).

Environmental awareness of the concepts of environmental education and the environment was assessed by means of questions, enabling free responses that allowed people to express their thoughts without being induced. However, it is extremely important to emphasise that we are not only concerned with working on concepts and definitions, but it is also necessary to develop actions in primary schools that value and highlight local social, environmental, political and cultural aspects, so that we can seek to train "environmental educators" and/or subjects who are critical and reflective of their role in society.

In view of all the questions that have been raised, in order for there to actually be changes, we must look for processes to educate students that can provide new perceptions of the environment, observing the possibility of a new pedagogical perspective for environmental education practices, thus realising that part of their education is being built towards an ecological individual.

The results obtained show a rapprochement in terms of the educational activities that can be developed in the park, as well as greater interaction between the school and the park. Including environmental education projects in the education of students can be a good step towards broadening the scope of activities carried out in natural areas, whether they are part of a protected area or not; however, with some caveats, any project already developed in schools about the Barão de Mauá Park needs to be better publicised among teachers, since they have little involvement with the area. EE is still an emerging area that is being developed in various sectors of society, and is of great importance for the training of students and professionals. This could lead to the development of natural resource management projects, favouring not only individual but also collective growth, with political, social, economic and environmental aspects being linked, taking into account local specificities.

The proposals below can be used as a suggestion to help outline ideas for working on Environmental Education pedagogical action in Nature Parks, strengthening their relationship with the students who live in their surroundings.

1) Extension of lessons to areas of the CU: use of interpretive trails as a teaching resource;

2) Development of educational activities and interventions that empower and stimulate students in the importance of being a participating individual;

3) Developing and making available teaching material so that students feel they are an integral part of learning, so that activities make sense, i.e. contain the peculiarities of the place.

This set of proposals can be seen as an increase in the educational process that aims to train participative individuals with the interventions that were outlined with them in this research. The reports made through the answers given by the students indicate that the trajectory of the work makes more sense when accompanied by constant questioning.

CHAPTER 9

REFERENCES

AFONSO, A. J. *Sociology of non-school education: reactualising an object or constructing a new problematic?* In: ESTEVES, Antônio J *A sociologia na escola: professores, educação e desenvolvimento.* Porto: Afrontamento. 1992. Library of the Science of Man.

ALDRIDGE, DON. Improving Park Interpretation and Communication with the Public, (ed.), *Second World Conference on National Parks',* Yellowstone Grand Teton, EE. U.S.A., 18-27 September 1972. Report No. 25.

AVILA, F. B. de. *Little Encyclopaedia of Morals and Civics.* Rio de Janeiro: FENAME, 1972. 698p.

BRANDÃO, C. R. *O que é educação.* 23 ed. São Paulo: Brasiliense, 116p.1981.

BRAZIL. Secretariat of Basic Education. *Parâmetros Curriculares Nacionais: Introdução aos Parâmetros Curriculares Nacionais/* Secretaria de Educação Fundamental- Brasília: MEC/SEF, 1997. PI26.

BRAZIL. MMA. *Law no. 9.795, of 27 April 1999.* Provides for environmental education, institutes the National Environmental Education Policy and makes other provisions. Brasília: MMA/SBF, 1999.

BRAZIL. MMA. *Law no. 9.985, of 18 July 2000.* Regulated by Decree 4.340 of 22/08/2002. National System of Nature Conservation Units - SNUC: Brasília: MMA/SBF, 2000.

BRAZIL. Ministry of Education and Sports. National Education Council. Full Council. *Resolution no. 2, of 15 June 2012.* Establishes the National Curriculum Guidelines for Environmental Education. *Federal Official Gazette,* Brasília, DF, 18 June 2012. Available   at: <http ://www.in.gov.br/visualiza/index.j sp?data=18/06/2012&jomal= 1 &pagina=71 &totalArquiv os=320>. Accessed on: 16 September 2015.

BRAZIL. MMA. National Environmental Education Programme (PRONEA). *Environmental Education for a Sustainable Brazil.* Brasília: MMA, 2014.

BERGER, J. *Modos de Ver.* Translation by Lúcia Olinto. Rio de Janeiro: Rocco, 1999.

BELART, J. L. *Trails for Brazil. Boletim FBCN,* Rio de Janeiro, v.13, n.l, p.49 -51 1978.

BENTO, J. O. In *defence of sport,* In Bento, J. O. and CONSTATINO, J. M. *In defence of sport: mutations and conflicting values.* Coimbra: Almandina, 2007.

BOGDAN, R. C. & BIKLEN, S. K. *Qualitative research in education: an introduction to the theory of methods.* Porto: Porto Editora, 1994, in *T* LIMA, Paulo Gomes. *Paradigmatic trends in educational research.* Campinas, SP: (s.n), 2001.

BOOTH, W.C.; COLOMB, G.G.; WILLIAMS, J.M. *The Art of Research.* São Paulo: Livraria Martins Fontes Editora Ltda., la ed., 2000.

CAMARA, I. de G. *Megabiodiversity.* Rio de Janeiro: Sextante, 2001.

CERVO, A. *M* BERVIAN, P. A. *Metodologia científica.* 5. ed. São Paulo: Makron Books, 2002

CORNELL, J. *The joy of learning with nature: outdoor activities for all ages.* Translation: Maria Emília de

Oliveira; São Paulo: Editora SENAC. 1997.

CORTELLA, M. S. *A contribuição da educação não formal para construção da cidadania;* Cortella, Mario S; SIIMSOM, Olga R. von; PARK, Margareth; FERNANDES, Renata et al. *Visões singulares, conversas plurais.* São Paulo: Instituto Itaú Cultural, 2007, p. 43-52.

CORTELLA, M. S. *Contributions of non-formal education to the construction of citizenship.* Lecture given at the Itaú Cultural Institute, 6 December 2007.

DIAS, F.V.; ZANIN, E.M. *Efficiency of interpretive trails in the Longines MMalinowski Municipal Park -* Erechim-RS. *Perspectiva,* v.28, p.29-38, 2004.

FERREIRA, C. P. *Environmental Perception in the Juréia-Itatins Ecological Station. Master's dissertation.* University of São Paulo, SP. 2005.

FREIRE, P. *Pedagogy of autonomy.* Petrópolis: Vozes, 2003.

FREITAS, F; MARTINS, I. P. *Promoting science learning in the 1st CEB using non-formal education contexts, Ensenanza de las ciencias,* n. extra, vii congresso, p. 1-4, 2005.

FOLENA, S.; ANJOS, M. *Pre-school education and the environment: a propositional discussion. Revista Educação Ambiental em Ação,* n. 13, 2005. Available at: <http://www.revistaea.org/artigo.php?idartigo=323&class=21>. Accessed on: 16 Apr. 2014.

GADOTTI, M. *Pedagogy of the Earth: Eco-pedagogy and sustainable education. Pedagogy of the Earth.* São Paulo: Petrópolis Foundation, 2005.

GIL, A. C. *Métodos e técnicas de pesquisa social.* SP, Atlas, 4ª ed., 1995.

GOHN, M. G. *Non-formal education, civil society participation and collegiate structures in schools.* Ensaio: aval. pol. publ. Educ., Rio de Janeiro, v.14, n.50, p.27- 38, jan./mar. 2010.

GUIMARÃES, S. T DE L. *Perception, interpretation and environmental education: a geographical perspective. Territory & Citizenship.* São Paulo, vol. 3, n.l, 2003. Available at:

<http://www.rc.unesp.br/igce/planejamento/territorioecidadania>. Accessed on: 20 November 2014.

HAM, S. *I Environmental Interpretation. A practical guide for people with big ideas and small budgets,* Forest Wildlife and Range Experiment Station, University of Idaho, USA. 437 pp. 1992.

JACOBUCCI, D. F. C. *Contributions of non-formal educational spaces to the formation of scientific culture. In extension,* Uberlândia, V.7, 2008.

JESUS, J .S.; RIBEIRO, E .M .S. *Diagnosis and proposal for implementing a trail in the Armando de Holanda Cavalcanti Metropolitan Park, Cabo de Santo Agostinho, PE.* In: *Proceedings of the 1st National Congress on Trail Planning and Management.* Rio de Janeiro: Infotrilhas, 2006.

KRASILCHIK, M. *Reforms and reality: the case of science teaching. São Paulo em Perspectiva* vol.14 n.l. São Paulo, p. 85-93. Jan./March. 2000.

LAKATOS, E. M; MARCONI, M. A. *Metodologia do Trabalho Científico: procedimentos básicos, pesquisa bibliográfica, projeto, relatório, publicações e trabalhos científicos.* 4 ed. São Paulo: Atlas, 1992.

LIBÂNEO, J. C. *Democratização da Escola Pública: a Pedagogia Critica Social dos Conteúdos.* 9. ed. São Paulo: Loyola, 2005.

LONDONO, J. L., GUERRERO, R. *Violence in Latin America: Epidemiology and Costs.*

Washington, D.C.: Inter-American Development Bank, Office of the Chief Economist, 1999.

LÚDKE, M: ANDRE, M. E. D. A. *Research in Education: Qualitative Approaches.* São Paulo: EPU, 2014.

MACHADO, L. M. C. P. **3º** *Interdisciplinary Meeting on the Study of Landscape.* Rio Claro: UNESP, 1998. v.l. 154 p. (Cadernos Paisagem/Paisagens).

MARCONI, M. A; LAKATOS, E. M. *Metodologia do Trabalho Científico.* São Paulo: Editora Atlas, 1992. 4th ed. p.43 and 44.

MARANDINO, M. *Da Ciência Biologia ao Ensino e Biologia nos Espaços Formal e Não- Formal.* In: Selles et al. *Proceedings of the II Regional Meeting on Biology Teaching* - Regional 02. Niterói, 2003.

MAURICE S. and MIKHAIL , G. *Earth Charter.* Members of the Commission to support the Earth Charter Initiative and act as Earth Charter Ambassadors. Following the launch of the Earth Charter in 2000, the Commission handed over responsibility for overseeing the Earth Charter Initiative and fundraising to a Steering Committee. In 2006, the Steering Committee was replaced by the CTI.1997 Board.

MELLO, F. A. P.; *Trail management: more than closing shortcuts and building steps, a transdisciplinary approach.* In: COSTA, N. M. C.; NEIMAN, Z.; COSTA, V. C. (Org.). *Pelas Trilhas do Ecoturismo.* São Carlos: RIMA, 2006, 187-200p.

MENGHINI, F. *Interpretive Trails as a Pedagogical Resource: paths mapped out for environmental education.* 2005. 103 f. Dissertation (Master's in Education) - Area of concentration: Education - (Research Line: Teacher Training and Professional Identities, Research Group, Education, Environmental Studies and Society/GEEAS), University for the Itajaí Valley, Itajaí, 2005.

MILANO, M. S. *Concepts and general principles of ecology and conservation. In:* FUNDAÇÃO O BOTICÁRIO DE PROTEÇÃO À NATUREZA (Ed.). Course on administration and management in Conservation Units. Curitiba: **FBPN,** 2001 p. 1-55.

MORALES, J: *Manual para la Interpretación Ambiental en Areas Silvestres Protegidas,* Documento Técnico # 8, proyecto **FAO/ PNUMA.** 201 pp.1992.

MOORE et al: *Manual para la capacitación del personal de áreas protegidas* (Módulo C : *Interpretación y Educación Ambiental,* Apunte 1b), Washington D. C.,USA : National Parks Service, Office of International Affairs. 1989.

MORIN, E. *The well-made head: Rethinking Reform, Reforming Thought.* Eloá Jacobina. Rio de Janeiro: Bertrand Brasil, 2000.

OLIVEIRA, C. L.; MOURA, D. G. *Methodology of projects and non-formal learning environments: Beginning of effectiveness in the process of teaching Biology.* **Proceedings of V ENPEC-** n° 5, 2-3 p. 2005.

SWEET FORESTS PROJECT. *Introduction to Environmental Interpretation.* IEF - IBAMA - Biodiversitas Foundation - GTZ. Belo Horizonte. 108p. 2002.

PINHEIRO, N. A. M. *Critical-Reflective Education for a Scientific-Technological High School: the Contribution of the CTS Approach to the Teaching-Learning of Mathematical Knowledge.* Thesis (Doctorate in Scientific and Technological Education). Federal University of Santa Catarina, Florianópolis, 2005.

REIGOTA, M. *O que é educação ambiental.* São Paulo: Brasiliense, 1998. *Electronic Magazine on Environmental Education in Action.* Interview with Genebaldo Freire Dias. n° 15, year IV, Dec-Feb 2006.

SANDIN E.M. P. *Pesquisa Qualitativa em Educação: fundamentos e tradições.* Porto Alegre: Artmed, 2010.

SHARPE, G. W. *Interpreting the environment.* CATIE, Costa Rica. pp. 2-3. 1988.

SAUVÉ, L. L. *Education Relative à l'environnement: une Dimension Essentielle de l 'Education Fondamentale.* In Gohier, Cristiane and Laurin, Suzanne (2001). La *formation fondamentale- Un espace a redefinir.* Montreal: Les Editions Logiques, p. 293- 318.2001.

SATO, M. *Education for the Environment.* PhD thesis presented to the Postgraduate Programme in Ecology and Natural Resources. Federal University of São Carlos. São Paulo, 1997.

SATO, M; SANTOS, J. E. *Trends in Environmental Education Research.* In: NOAL, F.; BARCELOS, V. (Orgs.) *Educação ambiental e cidadania: cenários brasileiros.* Santa Cruz do Sul: EDUNISC, 2001, p. 253-283.

SOUZA, F. A. de. *Environmental Education: A methodological proposal for primary and secondary education.* Cajazeiras (PB): Vitoriano, 2002. 90p.

SMITH, J, C. *Environmental education: a view of a changing scene.* In: *EnvironmentalEducation Research,* v. l,n. 1. 1995.

TILDEN, F. *Interpreting our Heritage.* 3ª edition. North Carolina: The University of North Carolina Press. 117p. 1977.

TIRIBA, L. *Economic science and popular knowledge: reclaiming the "popular" in economics and education.* In: TIRIBA, L.; PICANÇO, I. (eds): *Trabalho e Educação: arquitetos, abelhas* e *outros tecelões da economia popular solidária.* Ideias e Letras, 2004.

TRILLA, J. *Non-formal education.* In: ARANTES; Vaieria Amorim (Org.) *Educação Formal e não Formal.* São Paulo: Summus, 2008.

TRISTÃO, M. *Saberes e Fazeres da Educação Ambiental no Cotidiano Escolar.* In: *Revista brasileira de educação ambiental* / Rede Brasileira de Educação Ambiental. - n. 0 (nov.2002). - Brasília: Rede Brasileira de Educação Ambiental, 2004, 140 p. v. il.; 28 cm. 2002, Quarterly.

TUAN, Y. F. *Topophilia: A Study of Perception, Attitudes and Values of the Environment.* Ed. Difel, São Paulo. 1980.

VASCONCELLOS, J. *Interpretive Trails: Combining Education and Recreation.* In: Carlos, A.F. A e outros, (org.) Turismo, Espaço e Cultura. São Paulo: HUCITEC, 1997. 465 - 477 P.

VASCONCELLOS, J. M. O.OTA, S. *Ecological Activities and Interpretive Trail Planning.* Maringá: Agronomy Department, UEM, 2000 (mimeo).

VASCONCELLOS, J. M. O. *Evaluation of Public Visitation and the Efficiency of Different Types of Interpretive Trails in Pico do Marumbi State Park and Salto Morato* **Nature** *Reserve* - PR. Curitiba. 1998. 141fls. Thesis (Doctorate in Forestry Sciences). Postgraduate Programme in Forestry Engineering, Federal University of Paraná.

VIDAL, J. Um *Diálogo Entre a Política Cultural e a Educação não Formal: Contribuições para o Processo de Constituição da Cidadania das Pessoas com Deficiência,* 2009. Master's dissertation - Faculty of Education, University of São Paulo.

VIEIRA, L. R. S. *Environmental Awareness at the Saint Gobain Canalisation Machinery, Itaúna-MG: an EE methodology for industry.* São Carlos. Doctoral Thesis. São Carlos School of Engineering. University of São Paulo. 2005 143p

ZANIN, E. M. *Projeto Trilhas Interpretativas - a extensão, ensino e pesquisa integrados à conservação ambiental e à educação.* Vivências, Erexim. v.l, Anol, May, n^0 . 2, p. 2635.2006.

Comitê de Ética em Pesquisa

Neighbourhood where you live CITY

Gender Male( ) Female ( )

Date of Birth: _______//

Questionnaire

1-Is this the first time you've been hiking in the conservation unit you've visited?

( ) Yes   ( )No

2-   What do you expect to find on the Park trail? (If you prefer, use the back of the sheet to draw a picture).
____
____
____
____
____

3-   According to what you observed during your visit to the park, what do you consider to be an environmental problem? NOTE: Tick as many boxes as you like.

( ) Water pollution ( ) Waste ( ) Soil contamination ( ) Animal extinction.

( ) Extinction of plants ( ) Lack of green areas ( ) Maintenance of the conservation unit

4-   According to what has been observed, who is responsible for the degradation of the park: ( ) The government. ( ) The people. ( ) I don't know

5-   What suggestions would you make to reduce the degradation of the park?
____
____
____
____

6-   Did you enjoy the park's trails?
____
____
____

7- What activities can be carried out in the park:

( ) Leisure.   ( ) Environmental education. ( ) Tourism. ( )   No
you know.

44

8- You'd like to visit the park again.

( ) Yes ( ) No

ANNEX 2- LETTER OF CONSENT FROM THE HOST INSTITUTION.

CARTA DE ANUÊNCIA da INSTITUIÇÃO SEDIADORA

45

Declaramos, para os devidos fins, que concordamos em disponibilizar o(s) setor(es) _Ensino Fundamental (6° ao 8° ano)_ desta Instituição, para o desenvolvimento das atividades referentes ao Projeto de Pesquisa, intitulado: _TRILHAS INTERPRETATIVA NA EDUCAÇÃO AMBIENTAL_, dos pesquisadores _MARCO ANTONIO DA SILVA VIEIRA_ sob a responsabilidade do Professor _JOÃO RODRIGUES MIGUEL_ do curso de _MESTRADO_, da Universidade do Grande Rio, pelo período de execução previsto no referido Projeto.

Rio de Janeiro, _18_ de _02_ de _2014_

Angeli Máximo da Costa Macedo
Nome, por extenso, do responsável pelo setor
Diretora Adjunta
Cargo e/ou função que exerce na instituição
Augusto
Assinatura e Carimbo

_0194 7939/0001-20_
CPF / CNPJ

emphildaseoelho.mage@gmail.com
E-mail

Secretaria Municipal de Meio Ambiente
Prefeitura Municipal de Magé
Estado do Rio de Janeiro

CARTA DE ANUENCIA UC 001/2014 – Parque Natural Barão de Mauá

Declaro que a Secretaria Municipal de meio Ambiente de Magé – Órgão Gestor da Unidade de Conservação Municipal Parque Natural Barão de Mauá, não tem nada a opor a realização do Projeto de Pesquisa Trilhas Interpretativas, como conclusão de Mestrado do Biólogo Marco Antônio da Silva Vieira, do Curso Mestrado em Ensino das Ciências da Universidade do Grande Rio. O Projeto utilizará alunos da rede de ensino municipal e tem como objetivo fazer o reconhecimento do ecossistema local, visitação guiada e interpretativa das trilhas e do ambiente costeiro da Unidade de Conservação Municipal.

Sem mais nos colocamos a disposição para o fornecimento de mais informações, assim como no apoio costumeiro no que tange o desenvolvimento sustentável e a preservação ambiental.

Magé 18 de fevereiro de 2014

Leandro Vidal Santos
Secretário de Meio Ambiente
SMMA - Mat. 354577

Estrada das Margaridas, s/nº – Santa Dalila – 4º Distrito de Magé – RJ – CEP 25.925-000
Tel.: (21) 2647-1214 – E-mail: meioambiente@mage.rj.gov.br

COMITÊ DE ÉTICA EM PESQUISA

Duque de Caxias, 17 de Novembro de 2014.

Do: Comitê de Ética em Pesquisa da UNIGRANRIO

Para Pesquisador: Marco Antonio da Silva Vieira

Orientador: Prof. Dr. João Rodrigues Miguel

O Comitê de Ética em Pesquisa da UNIGRANRIO, após avaliação considerou **aprovado** o projeto de pesquisa **"TRILHAS INTERPRETATIVAS NA EDUCAÇÃO AMBIENTAL"**, protocolado sob o **número de CAAE. 27978914.5.0000.5283**, encontrando-se a referida pesquisa e o Termo de consentimento Livre e Esclarecido em conformidade com a Resolução N.º 196, de 10 de outubro de 1996, do Conselho Nacional de Saúde, sobre pesquisa envolvendo seres humanos.

Os pesquisadores deverão informar ao Comitê de Ética qualquer acontecimento ocorrido no decorrer da pesquisa.

O Comitê de Ética em Pesquisa solicita a V. Sª., que ao término da pesquisa, conforme cronograma apresentado, encaminhe a este comitê um sumário dos resultados do projeto, a fim de que seja expedido o certificado de aprovação final.

Prof. Renato C. Zambrotti
Coordenador do CEP-UNIGRANRIO

Andreia Peter Christo Gomes
Secretária do CEP/UNIGRANRIO

CEP/UNIGRANRIO – COMITÊ DE ÉTICA EM PESQUISA da UNIGRANRIO
Rua Prof. José de Souza Herdy, 1160 – 25 de Agosto – Duque de Caxias – CEP: 25071-202
Tel.: 21 2672-7733 – E-mail: rzambrotti@unigranrio.com.br

4 **MAGÉ** **BOLETIM INFORMATIVO OFICIAL** 16 A 31 DE OUTUBRO DE 2012 EDIÇÃO 431

ANEXO I – MAPA DA UNIDADE PROPOSTA

MAGÉ REPÚBLICA FEDERATIVA DO BRASIL - ESTADO DO RIO DE JANEIRO SECRETARIA MUNICIPAL DE ADMINISTRAÇÃO - www.mage-rj.gov.br

PORTARIAS

PORTARIA Nº. 1261/2012.
O PREFEITO DO MUNICÍPIO DE MAGÉ, no uso de suas atribuições legais e em conformidade com o art.71 da Lei Orgânica Municipal, e,
CONSIDERANDO, as informações constantes no Processo nº.3504/2012 e parecer da Procuradoria Geral do Município.
RESOLVE:
INSTAURAR, SINDICÂNCIA ADMINISTRATIVA para no prazo de 30(trinta) dias, apurar o retorno ao serviço da servidora SONIA REGINA DOS SANTOS sem o competente ato administrativo que determinasse sua readmissão, ficando a cargo da Comissão constituída através da Portaria nº.068/2012 a adoção das medidas cabíveis.
PREFEITURA MUNICIPAL DE MAGÉ, EM 20 DE SETEMBRO DE 2012.
NESTOR DE MORAES VIDAL NETO
PREFEITO
OMITIDO NO BIO DA 2ª QUINZENA DE SETEMBRO DE 2012.

PORTARIA Nº. 1306/2012.
O PREFEITO DO MUNICÍPIO DE MAGÉ, no uso de suas atribuições legais e considerando o artigo 91, inciso II, alínea "a", da Lei Orgânica do Município,
RESOLVE:
CONCEDER, conforme requerimento protocolado sob o nº. 22.609/2012 e de acordo com o art.128, parágrafo único da Lei Municipal nº.1054/991, LICENÇA ESPECIAL ao servidor JOSÉ ROBERTO ABELLEIRA PIERASSOL DOS SANTOS, matrícula T-0712 – Agente Administrativo III da Secretaria Municipal de Administração, pelo período de 03(três) meses relativo ao quinquênio 2000/2005, com efeito a 05 de agosto de 2012 e término em 05 de novembro de 2013.
PREFEITURA MUNICIPAL DE MAGÉ, EM 26 DE SETEMBRO DE 2012.
NESTOR DE MORAES VIDAL NETO
PREFEITO
OMITIDO NO BIO DA 2ª QUINZENA DE SETEMBRO DE 2012.

PORTARIA Nº.1311/2012.
O PREFEITO DO MUNICÍPIO DE MAGÉ, no uso de suas atribuições legais e considerando o artigo 91, inciso II, alínea "a", da Lei Orgânica do Município, e
Considerando o interesse municipal na Reativação da Estrada de Ferro Barão de Mauá.
RESOLVE:
CRIAR, GRUPO DE TRABALHO-TREM/MAGÉ, constituído dos seguintes Membros:
- ANTONIO CARLOS MERITELLO MACHADO – Jornal Bate Papo;
- HÉLIO SUÉVO RODRIGUEZ – AMUTREM (Amigos Museu do Trem-RJ.);
- LUIZ OCTAVIO OLIVEIRA – Vice-Presidente da AFPF (Associação Fluminense de Preservação da Ferrovia);
- ANTONIO PASTORI – Coordenador do GFPF (Grupo Fluminense de Preservação Ferroviária);
- RUBEM EDUARDO LADEIRA – Diretor da AENFER (Associação de Engenharia Ferroviária;
- EDIR INACIO DA SILVA – Secretário de Trabalho, Emprego, Habitação e Geração de Renda do Município de Magé.
PREFEITURA MUNICIPAL DE MAGÉ, EM 02 DE OUTUBRO DE 2012.
NESTOR DE MORAES VIDAL NETO
PREFEITO
OMITIDO NO BIO DA 1ª QUINZENA DE OUTUBRO DE 2012.

PORTARIA Nº. 1322/2012.
O PREFEITO DO MUNICÍPIO DE MAGÉ, no uso de suas atribuições legais, e em conformidade com o art. 91, inciso II, alínea "a" da Lei Orgânica Municipal,
RESOLVE:
NOMEAR, o Sr. ANTONIO SERGIO DE O. AYMORE MARTINS, para ocupar o Cargo em Comissão de SUPERVISOR DE TRABALHO, índice DA-3, da Secretaria Municipal de Trabalho, Emprego, Habitação e Geração de Renda, com efeito a 10 de outubro de 2012.
PREFEITURA MUNICIPAL DE MAGÉ, EM 16 DE OUTUBRO DE 2012.
NESTOR DE MORAES VIDAL NETO
PREFEITO

PORTARIA Nº. 1325/2012.
O PREFEITO DO MUNICÍPIO DE MAGÉ, no uso de suas atribuições legais, e em conformidade com o art. 91, inciso II, alínea "a" da Lei Orgânica Municipal,
RESOLVE:
NOMEAR, o Sr. LUIS HENRIQUE DOS SANTOS TEIXEIRA, para ocupar o Cargo em Comissão de DIRETOR DE AGRICULTURA, índice DA-2, da Secretaria Municipal de Desenvolvimento Econômico e Agricultura, com efeito a 10 de outubro de 2012.
PREFEITURA MUNICIPAL DE MAGÉ, EM 16 DE OUTUBRO DE 2012.
NESTOR DE MORAES VIDAL NETO
PREFEITO

2 MAGÉ **BOLETIM INFORMATIVO OFICIAL** 16 A 31 DE OUTUBRO DE 2012  EDIÇÃO 431

MAGÉ
REPÚBLICA FEDERATIVA DO BRASIL · ESTADO DO RIO DE JANEIRO
SECRETARIA MUNICIPAL DE ADMINISTRAÇÃO · www.mage.rj.gov.br

DECRETOS

DECRETO Nº 2.793/2012
O PREFEITO MUNICIPAL DE MAGÉ, no uso de suas atribuições legais e com base no art. 5º da Lei 2.171 de 26/07/2012;
Considerando, ainda, a indispensável adequação das dotações orçamentárias de diversas Unidades Orçamentárias, face as sua necessidade e atribuições,
DECRETA:
Art. 1º - Ficam suplementados os elementos de despesa, das atividades constantes dos Programas de Trabalho de governo, conforme discriminado no anexo I, parte "A", parte integrante do presente decreto, totalizando a importância de R$ 7.837.088,27 (Sete milhões oitocentos e trinta e sete mil e oitenta e oito reais e vinte e sete centavos).
Art. 2º - Os recursos necessários para fazer face à suplementação de que trata o Art. 1º, provêm da anulação parcial no elemento de despesa do projeto e atividade constante do Programa de Trabalho de Governo, conforme discriminado no anexo I, parte "B", parte integrante do presente decreto no valor de R$ 7.837.088,27 (Sete milhões oitocentos e trinta e sete mil e oitenta e oito reais e vinte e sete centavos).
Art. 3º - Em decorrência dos dispostos nos artigos 1º e 2º, fica alterado o Quadro de Detalhamento de despesa do Fundo Municipal de Desenvolvimento e Assistência Social, Fundo Municipal de Saúde, Prefeitura Municipal de Magé e Fundo Municipal de Habitação.
Art. 4º - Este decreto entrará em vigor nesta data, revogadas as disposições em contrário.
Magé, 01 de outubro de 2012.
NESTOR DE MORAES VIDAL NETO
Prefeito
OMITIDO NO BIO DA 1ª QUINZENA DE OUTUBRO DE 2012.

ANEXO I - DECRETO Nº 2.793 de 01/10/2012 - PARTE "A" — SUPLEMENTAÇÃO

PROGRAMA DE TRABALHO	CÓDIGO DESPESA	ELEMENTO DESPESA	FONTE DE RECURSOS	VALOR
04.122.0012-2.004	2	3.1.90.11.01	100	224.465,21
04.122.0012-2.006	10	3.1.90.11.01	100	76.234,98
04.122.0012-2.009	20	3.1.90.11.01	100	388.446,83
04.122.0015-2.011	28	3.1.90.11.01	100	33.007,00
04.122.0012-2.013	39	3.1.90.11.01	100	131.708,12
08.122.0020-2.019	61	3.1.90.11.01	100	106.827,50
15.122.0006-2.107	86	3.1.90.11.01	100	1.241.061,35
04.122.0014-2.021	106	3.1.90.11.01	100	332.819,00
04.122.0015-2.026	120	3.1.90.11.01	100	384.338,71
04.122.0037-2.074	183	3.1.90.11.01	100	86.021,00
23.695.0031-2.076	190	3.1.90.11.01	100	65.034,00
04.122.0037-2.078	191	3.1.92.11.01	100	35.335,69
12.361.0027-2.063	251	3.1.90.11.01	412	1.572.098,99
08.122.0093-2.045	332	3.1.90.11.01	100	319.674,80
15.452.0000-2.117	325	3.1.90.11.01	100	141.388,94
26.782.0020-2.123	403	3.1.90.11.01	100	60.529,67
16.122.0004-2.124	422	3.1.90.11.01	100	56.645,29
04.121.0005-2.0121	433	3.1.90.11.01	100	16.035,00
11.122.0012-2.138	440	3.1.90.11.01	100	19.677,00
12.365.0057-1.068	504	3.1.90.04.00	407	30.900,00
08.122.0003-2.045	332	3.1.90.11.01	100	20.788,00
04.122.0013-2.013	38	3.1.90.09.00	100	3.000,00
12.361.0027-2.071	608	3.3.90.93.00	410	15.001,18
08.243.0023-2.155	888	3.1.90.11.01	211	3.375,00
08.244.0023-2.092	005	3.3.90.93.00	12	2.972,12
10.302.0019-2.150	607	4.5.90.61.00	110	75.000,00
10.302.0019-2.063	621	3.3.90.39.00	521	26.703,56
10.302.0019-2.150	624	3.3.90.39.00	502	327.000,00
10.301.0004-2.005	651	3.3.20.93.00	510	219.554,66
10.302.0019-1.027	902	4.4.90.52.00	110	528.192,50
10.302.0019-2.160	624	3.3.90.39.00	502	698.000,00
10.301.0019-2.151	660	3.3.90.39.00	505	401.000,00
10.301.0004-2.005	650	3.3.20.93.00	501	18.000,00
TOTAL				**7.837.088,27**

ANEXO I - DECRETO Nº 2.793/2012 de 01/10/2012 - PARTE "B" — ANULAÇÃO

PROGRAMA DE TRABALHO	CÓDIGO DESPESA	ELEMENTO DESPESA	FONTE DE RECURSOS	VALOR
12.361.0027-2.064	259	3.1.90.11.01	412	1.572.098,99
12.361.0027-2.065	273	3.1.90.11.01	400	3.730.904,24
12.365.0057-2.1.068	326	3.3.90.39.00	407	30.900,00
12.361.0027-2.065	273	3.1.90.11.01	400	26.700,00
12.361.0027-2.157	514	3.3.90.30.00	410	15.001,18
08.243.0023-2.155	673	3.3.90.04.00	211	3.375,00
08.244.0003-2.068	690	3.3.90.39.00	201	672,12
08.244.0023-2.092	707	3.3.90.93.00	210	1.000,00
08.244.0023-2.092	708	4.4.20.93.00	210	1.000,00
16.482.0037-1.052	812	4.4.90.51.00	110	600.192,50
10.302.0019-2.093	614	3.3.90.30.96	521	26.703,56
10.302.0019-2.150	518	3.3.90.30.09	502	177.000,00
10.302.0019-2.150	521	3.3.90.30.36	502	150.000,00
10.302.0019-2.034	502	3.1.90.14.00	501	218.554,66
10.302.0019-1.027	498	4.4.90.52.00	502	150.000,00
10.301.0019-2.151	547	3.3.90.30.36	505	401.000,00
10.301.0004-2.005	555	3.3.90.92.00	501	18.000,00
10.302.0019-2.160	518	3.3.90.30.09	502	718.000,00
TOTAL DAS ANULAÇÕES				**7.837.088,27**

DECRETO Nº 2795/2012
Dispõe sobre a criação do PARQUE NATURAL MUNICIPAL BARÃO DE MAUÁ.
O PREFEITO DO MUNICÍPIO DE MAGÉ, no uso de suas atribuições legais, com base no que dispõe o Artigo 68, inciso IV e VIII, da Lei Orgânica do Município, e tendo em vista o disposto nos arts. 11 e 22 da Lei nº 9.985, de 18 de julho de 2000, regulamentados pelo Decreto nº 4.340, de 22 de agosto de 2002, e o que consta do processo nº 04-004439/2012.
DECRETA:
Art. 1º - Fica criado o PARQUE NATURAL MUNICIPAL BARÃO DE MAUÁ, com área [illegible], no Município de Magé, 5º Distrito Guia de Pacobaíba, Ipiranga.
§ 1º - O memorial descritivo dos limites do parque consta do Anexo I do presente Decreto.
§ 2º - O mapa de situação do parque consta no Anexo II do presente Decreto.
Art. 2º - A criação do Parque Natural Municipal Barão de Mauá tem por objetivos:
I- Preservar e recuperar as áreas degradadas existentes do ecossistema do manguezal e a conservação da biodiversidade associada ao bioma da Mata Atlântica;
II- Realizar pesquisas científicas;
III- Desenvolver atividades de visitação, recreação, educação e interpretação ambiental, estimulando o desenvolvimento do turismo em bases sustentáveis;
IV- Proteger e preservar populações de animais e plantas nativas e oferecer refúgio para espécies migratórias, raras, vulneráveis, endêmicas e ameaçadas de extinção de fauna e flora nativas;
V- Assegurar a continuidade dos serviços ambientais prestados pela natureza.
Art. 3º - Fica estabelecida como de utilidade pública, para fins de desapropriação e implantação do Parque Natural Municipal Barão de Mauá, a área delimitada por este Decreto, sendo vedados empreendimentos, obras e quaisquer atividades que afetem sua subsistência ou destinação na área delimitada.
Art. 4º - O Parque será regido pela Lei Federal nº 9.985, de 18 de julho de 2000, que institui o Sistema Nacional de Unidades de Conservação da Natureza e pela legislação estadual e municipal pertinente.
Art. 5º - O Parque Natural Municipal Barão de Mauá será administrado pela Secretaria Municipal de Meio Ambiente, que adotará as medidas necessárias para sua efetiva implantação.
Art. 6º - Fica estabelecido o prazo máximo de 05 (cinco) anos, a partir da data de publicação deste Decreto, para a elaboração do plano de manejo do Parque Natural Municipal Barão de Mauá.
Art. 7º - Ficam permitidas na Zona de amortecimento do Parque Natural Municipal Barão de Mauá as atividades minerais autorizadas pelo Departamento Nacional de Produção Mineral e licenciadas pelo órgão ambiental competente até a data de publicação deste Decreto.
Art. 8º - Poderão ser permitidos, dentro dos limites da zona de amortecimento do Parque Natural Municipal Barão de Mauá, empreendimentos minerários, de exploração, produção, transporte dutoviário de petróleo e gás natural, de transmissão de energia elétrica, bem como demais atividades industriais que obtiverem as autorizações e licenças previstas na legislação, observadas as disposições do plano de manejo da unidade, quando houver.

PODER EXECUTIVO

Prefeito
NESTOR DE MORAES VIDAL NETO

Vice-Prefeito
CLÁUDIO FERREIRA RODRIGUES

Procurador Geral do Município
ALEX KLYEMANN BEZERRA PORTO DE FARIAS

Chefe de Gabinete
PAULO CÉSAR BATISTA VAZ

Secretaria de Governo
CREZIO DA SILVA SANTIAGO (interino)

Secretaria de Administração
ROBSON PEREIRA DE MELLO

Secretaria de Trabalho, Emprego, Habitação e Geração de Renda
EDIR INÁCIO DA SILVA

Secretaria de Fazenda
DIMAS DE ANDRADE PINTO

Secretaria de Planejamento
GUSTAVO MENEZES MORGADO (interino)

Secretaria de Transportes
RONALDO DA SILVA GONÇALVES (interino)

Secretaria de Controle Interno
OTTONI LUIZ FERREIRA

Secretaria de Obras
ROBSON ROBERTO DE ABREU

Secretaria de Meio Ambiente
LEANDRO VIDAL SANTOS (interino)

Secretaria de Segurança Pública
SAMUEL DIAS DIONÍZIO

Secretaria de Serviços Públicos
JOÃO CARLOS DA SILVA

Secretaria de Educação e Cultura
SANDRA MARA DE SOUZA BRITO

Secretaria de Esporte, Turismo, Lazer e Terceira Idade
ALEXANDRE BENTO RANGEL PINTO

Secretaria de Saúde
ROBERTO DAIUB ALEXANDRE (interino)

Secretaria de Assistência Social e Direitos Humanos
SELMA VAZ VIDAL

Secretaria de Manutenção Pública
EDIVAR SOUZA TAVARES

Secretaria de Habitação e Urbanismo
GUSTAVO MENEZES MORGADO

Secretaria de Desenvolvimento Econômico e Agricultura
ALOÍSIO PINTO STURM

Presidente da FECM
ALCILIA BRANDÃO TEIXEIRA

CÂMARA MUNICIPAL DE MAGÉ
TELS.: 2633-1605/3658-5615/2633-0547
ANDERSON COZZOLINO PRESIDENTE
Leonardo Franco Pereira - 1º Vice-Presidente
Sérgio Renato Pereira - 2º Vice-Presidente
Werner Benites Saraiva da Fonseca - 1º Secretário
Carlos da Silva Ferreira - 2º Secretário
Amisterdan Santos Viana
Álvaro Alencar
Guilherme Marcatti
Genivaldo Ferreira Nogueira
José Cortes Preto
Leandro Hessen Dim Rodrigues
Rafael Santos de Souza
Paulo Roberto Portugal

Interpretive trails in environmental education: Paths for raising environmental awareness

Marco Antônio da Silva Vieira
João Rodrigues Miguel

Summary

Based on the bibliographical studies of the authors that will be discussed and through the participatory observation that will be carried out, we will be able to verify the importance and use of these non-formal spaces by teachers for meaningful learning for themselves and their students. The well-planned, strategic/methodological non-formal space will provide students with the ideal place to complement their learning outside the formal space, through contact with living biodiversity. The use of the trail as a means of consolidating learning in science teaching in non-formal spaces is of great importance, because this trail leads the student, with the help of the teacher, to seek out knowledge that was previously unknown or avoided, because through contact with nature (fauna, flora, soil, air, water, etc.) and other environments available in each space visited, it will generate sensations and emotions in the students that would not normally manifest themselves during formal theoretical lessons.

Keywords: Interpretive trails. Environmental education. Environmental awareness

Certifico para os devidos fins que **MARCO ANTONIO DA SILVA VIEIRA** apresentou o trabalho **TRILHAS**

INTERPRETATIVA NA EDUCAÇÃO AMBIENTAL no I Encontro de Pesquisa em Ensino de Ciências e

Matemática: questões atuais, promovido pelo Programa de Pós-Graduação em Ensino das Ciências na

Educação Básica da UNIGRANRIO, realizado em Duque de Caxias no dia 03 de dezembro de 2013.

Duque de Caxias, 03 de dezembro de 2013

Prof. Dr. Carlos Henrique de Freitas Burity
Coordenador
Programa de Pós-Graduação em Ensino das Ciências na Educação Básica

MAGÉ MUNICIPAL NATURAL PARK AS A
TEACHING RESOURCE
FOR ENVIRONMENTAL AWARENESS

Student: MARCO ANTÔNIO DA SILVA VIEIRA
Collaborators:
Supervisor:PROF. PROF. JOÃO RODRIGUES MIGUEL
Course / Higher Education Institution: Postgraduate Diploma in Science Teaching / UNIGRANRIO

Introduction / Objectives / Material and Methods / Conclusions / Results / References

Krasilchik (1986) makes a brilliant reflection on the role of Environmental Education (EE): it must function as an integrating element, so that the community becomes aware of its development and its environmental implications, in addition to transmitting knowledge, it must develop skills and attitudes that allow man to act effectively in the process of maintaining environmental balance in order to guarantee a quality of life consistent with his needs and aspirations; for (EE) to fully achieve its objectives, some aspects must be considered: providing students with a solid base of knowledge, however, knowledge alone is insufficient, the basis of (EE) lies in the formation of a new awareness and participation (MIGUEL, NUNES & JASCONE, 2014). The aim of this work is to select, apply and evaluate group dynamics with the aim of legitimising ecology content. With the intention of "awakening student interest", the use of interpretive trails was adopted as a pedagogical resource for Science Teaching; the research involved 37 students from the 7th year and 30 from the 8th° year of primary schools II, from a Municipal school in Magé in the Baixada Fluminense; the out-of-class activity was carried out in the Barão de Mauá Municipal Natural Park, created by Municipal Decree 2.795/201, the total area of this Conservation Unit is approximately 116.80 hectares entirely covered by the mangrove ecosystem, located at the bottom of Guanabara Bay, with the mouth of the Estrela River, the Estrela Environmental Protection Area, the water mirror of Guanabara Bay and the dry land and urban areas surrounding the mangrove as its boundaries; it is located in the same city as the school, in the Ipiranga neighbourhood; data was collected through questionnaires, semi-structured interviews, analysis of official documents, dissertations and theses. School education in non-formal spaces is more comprehensive as it allows the student to be an integral part of the pedagogical process; the students were sensitised by the amount of "rubbish" that reaches the conservation unit via Guanabara Bay; a reality that filled these visitors with pride was getting to know the reality of this (UC) which had its area totally devastated and today, as a result of the activity of the NGO "Mangue Vivo", has been reforested with native mangrove species and the flora is also being recovered. The fact that the (UC) is part of the student's daily life, i.e. it is located close to their home, creates a responsibility to participate in the solution of the problems faced by the formation of an environmental conscience.

TERMO DE AUTORIZAÇÃO DE USO DE IMAGEM E DEPOIMENTOS

Eu _Eli Fortunato da Silva_

,CPF 437400361-2, RG 0 5443002-5. O seu filho ou o menor que você é responsável, está sendo convidado (a) a participar do projeto de pesquisa: **Trilhas Interpretativas na Educação Ambiental: Caminhos Para Sensibilização Ambiental. Na Unidade de Conservação Parque Natural Municipal Barão de Mauá.** A participação de seu filho neste projeto será de muita importância para nós, mas caso o mesmo desista de participar a qualquer momento isso não causará nenhum prejuízo ao seu filho ou a você como responsável. Ao mesmo tempo, libero a utilização de fotos (seus respectivos negativos) e/ou depoimentos para fins científicos e de estudos (livros, artigos, slides e transparências), em favor dos pesquisadores da pesquisa, especificados, obedecendo ao que está previsto nas Leis que resguardam os direitos das crianças e adolescentes (Estatuto da Criança e do Adolescente – ECA, Lei N.º 8.069/ 1990), dos idosos (Estatuto do Idoso, Lei N.º 10.741/2003) e das pessoas com deficiência (Decreto Nº 3.298/1999, alterado pelo Decreto Nº 5.296/2004).

Magé- RJ, _15_ de _maio_ de 2014.

Eli Fortunato da Silva
Marco Antônio da Silva Vieira
Pesquisador responsável pelo projeto

Vanina Ferreira Fortunato da Silva
Sujeito da Pesquisa

Printed by Books on Demand GmbH, Norderstedt / Germany